AUTUMN LAND-BIRD MIGRATION ON THE BARRIER ISLANDS OF NORTHEASTERN NORTH CAROLINA

PAUL W. SYKES JR.

Occasional Papers of the
North Carolina Biological Survey
1986-11

NORTH CAROLINA STATE MUSEUM OF NATURAL SCIENCES

PUBLICATIONS OF THE NORTH CAROLINA BIOLOGICAL SURVEY

The occasional publications of the North Carolina Biological Survey are devoted to scientific papers in various disciplines within the general fields of botany and zoology. Publications to be issued at irregular intervals include collections of short papers, book-length studies, and proceedings of symposia sponsored by the Survey.

Biological Survey publications are distributed by the North Carolina State Museum of Natural Sciences, P.O. Box 27647, Raleigh, N.C. 27611. Checks should be made payable to the NCDA Museum Extension Fund.

ELOISE F. POTTER
Editor

Occasional Papers of the
North Carolina Biological Survey
1986-11
$5, postpaid

Referees: Sidney A. Gauthreaux Jr. and Harry E. LeGrand Jr.

ISBN 0-917134125

AUTUMN LAND-BIRD MIGRATION ON THE BARRIER ISLANDS OF NORTHEASTERN NORTH CAROLINA

PAUL W. SYKES JR.

**Occasional Papers of the
North Carolina Biological Survey
1986-11**

DEDICATED

to

my parents, Paul W. Sykes Sr. and Alberta Carmine Sykes,

to my wife, Joan James Sykes,

and

to Thomas L. Quay

AUTUMN LAND-BIRD MIGRATION ON THE BARRIER ISLANDS OF NORTHEASTERN NORTH CAROLINA

Table of Contents

	Page
ABSTRACT	1
INTRODUCTION	1
METHODS	3
THE STUDY AREA	7
Major Habitats	8
Beach	8
Dune	9
Herb-shrub	9
Shrub Thicket	10
Pine Thicket	10
Fresh Marsh	10
High Marsh	10
Low Marsh	11
Climate	12
RESULTS	13
Magnitude and Duration of the Migration	13
Sequence of the Migration	22
Bird Movement Relative to Surface Weather	22
General Habitat Usage	32
DISCUSSION	32
ACKNOWLEDGMENTS	37
LITERATURE CITED	37
APPENDICES	41
Appendix A. Orders and Families of Birds Recorded in 1965	41
Appendix B. List of Birds Recorded in the Study Area in 1965	42
Appendix C. Ranking of Species by the Total Number of Individuals Recorded in 1965 by Censusing, Banding, and Collecting	46
Appendix D. Documentation of Accidental Species Recorded in the Study Area in 1965	48

AUTUMN LAND-BIRD MIGRATION ON THE BARRIER ISLANDS OF NORTHEASTERN NORTH CAROLINA

ABSTRACT.—An investigation in the autumn on the land birds occurring on the barrier islands in the Bodie Island and Pea Island area of northeastern North Carolina was initiated in 1964 and completed in 1966, with the major part of the study taking place during a 102-day period in 1965. The species richness, magnitude, duration, and sequence of migration were determined. Fifteen census areas were covered daily by walking line transects that included all major habitats, and a banding station was operated at intervals on Bodie Island. A total of 110,482 individuals of 148 species of birds representing 27 Families of 8 Orders were recorded in 1965. The Tree Swallow, Yellow-rumped Warbler, and Red-winged Blackbird were the three most abundant species. On the censuses, 53.8% of the species were migrants, 17.2% winter residents and migrants, 11.7% permanent residents and migrants, 6.9% summer residents and migrants, 5.5% permanent residents, 3.5% accidentals (= vagrants), and 1.4% summer residents. Of the 148 species recorded, 11% were abundant, 13% common, 20% fairly common, 26% uncommon, 26% rare, and 4% accidental. A total of 1,276 individuals (33% adults, 67% immatures) of 60 species were banded during 1,317 net hours on 14 days from 16 September through 9 November. Twenty-nine species were collected, 10 in 1965 and 23 in 1966.

The fall migration extended from mid-July through the last of November, and probably into mid-December. In general, the taxa of Neotropical affinity (i.e. hummingbird, flycatchers, and orioles) tended to migrate earlier in the fall than did those of Neartic affinity (i.e. wood warblers and sparrows). The heaviest movement was from mid-September through the last of October, with the peak in early October. The largest movements of birds were noted following the passage of strong cold fronts, usually with westerly winds. Some movement was also noted with the passage of weak fronts, and in a few cases there was some movement when there was no frontal passage. The small land mass and relatively limited amount of upland habitats on the barrier islands tended to concentrate the migrants. Although the study area is within the migration corridor of nearly all of the eastern North American land birds, many species tended to avoid the barrier islands and apparently reached the area mostly as a result of being drifted eastward from the mainland by strong westerly winds. This is supported by the fact that of the 148 species recorded in 1965, 57% were either uncommon, rare, or accidental in the area. The magnitude of the migration through the region was comparable with results along the Atlantic coast to the north. Five species native to the western United States were recorded as accidentals; three of these were the first reports for North Carolina and another species was only the second record for the state. The mechanism by which such vagrants may have reached the Atlantic coast is presented.

The phenomenon of bird migration has been recorded since the time of the ancients (Lincoln 1950). Although many studies on the causes and processes of migration have been done, migration is still not fully understood. Migration must be studied by both direct and indirect methods and in both the laboratory and the field (Pettingill 1956). In taking a census of grounded migrants one must realize that diurnal observations do not always reflect the volume of migration at a given locality for the previous night (H. E. LeGrand 1981). The number of nocturnal migrants observed in an area is not necessarily proportional to the number that passed overhead, as shown by comparing observations on radar at night with visual observations in the field the following day (Lack 1960, Drury et al. 1961).

In general, most workers agree that weather factors influence bird migration (Lack 1960; Richardson 1972, 1978). The important physical properties of the atmosphere are continually changing, resulting in complex meteorological conditions. The normal course of events in temperate latitudes alternates between anticyclonic and cyclonic systems. The movements of migrating birds show a strong correlation with factors associated with these conditions (Pettingill 1956). Bennett (1952) found that movement in the fall commenced immediately following the passage of a cold front. In North America in the fall, the majority of the great polar air masses move from northwest to southeast across the continent. The passage of these cold fronts is accompanied by a drop in temperature and clockwise shift of the winds to a northerly or northwesterly direction. Polar air masses are usually unstable, and as they move to the coast they absorb heat and moisture. Thus precipitation frequently accompanies the passage of a cold front. Studies of migration reveal that the weather factors that most consistently influence the density of fall migration are following winds and lower temperatures (Lack 1960, 1963; Gauthreaux 1978; Richardson 1978).

A significant amount of work has been done on fall migration of land birds in Europe and North America. In the United States most such studies have been in the Midwest and Northeast. The heavy concentrations of fall migrants along the northeast coast of the United States have been well documented (Allen and Peterson 1936, Stone 1937, Dennis and Whittles 1955, Baird et al. 1958, Baird and Nisbet 1959 and 1960). The first large-scale attempt to study land-bird migration along the Atlantic coast of North America was Operation Recovery from 1955 through 1969; it extended from southern Canada to the West Indies (Baird et al. 1958; U.S. Fish and Wildlife Service 1965; Chandler S. Robbins, pers. comm.). Operation Recovery spanned the months of August, September, and October; and since it terminated, a number of key banding stations have continued to operate in the fall. However, there has been very little study of migration on the coast south of Virginia.

Bishop (1901), Johnson et al. (1917). Burleigh (1937), and Duvall (1937) visited the North Carolina barrier islands at various seasons to study the avifauna. Nonetheless, the first significant work on the birds of this area was that of Green (1939). From May 1935 to February 1938, Green kept records of the birds observed from Hatteras Inlet north to Oregon Inlet, with the most intensive coverage in the vicinity of Buxton. He listed 84 land-bird species, most of which he observed during the 3-year period. Craighill and Grey (1938) visited that portion of the area from the Virginia-North Carolina state line south to and including Pea Island, and also Roanoke Island, from 29 September to 1 October 1937, and made some observations of the land birds. A number of short trips to the barrier islands were reported during the next 3 years by: Quay and Quay (1939) on 16 and 18 August 1939, Manteo to Pea Island; Craighill (1939), 8 to 13 September 1939, the Nags Head area; and Craighill (1940), early September 1940, Pea Island. The following year Grey (1941) published "The Breeding Birds of Pea Island," listing 13 land-bird species.

Commenting on the status of the wood warblers in North Carolina, Brimley (1942) reported that only four species of transient warblers had been noted east of Rocky Mount, these being the Blackpoll Warbler, the Black-throated Blue Warbler, the Northern Waterthrush, and the Palm Warbler. Further, he stated, ". . . there is certainly a coastal flyway along our beaches and 'banks', followed in the main by our migratory waterfowl and shore birds, and at least in the fall by the Tree and Barn Swallows and perhaps Purple Martins . . ." (*ibid.*, p. 41). Thus, little work had been done, prior to 1942, during migration along this section of the coast.

From 1949 to 1952, Crosson and Stevenson (1956) observed the hawk migration along the coast from Delaware to North Carolina and published a brief account in *The Chat.* Willoughby (1951a, 1951b) reported on his sightings at Hatteras, for the periods 21 to 28 August 1950 and 28 August to 4 September 1951. A number of land-bird migrants were found on each trip. From 17 to 23 September 1950, a few migrants were seen at Nags Head by Deignan (1951). Simpson (1954), reporting on the status of migratory hawks in the Carolinas, stressed the need for systematic observations on the coastal islands.

Quay (1959) conducted a survey of the birds of the Cape Hatteras National Seashore as a part of his study of the land vertebrates of the area. Field studies began on 8 November 1956 and were terminated on 10 May 1959. Trips were made during all four seasons, but most of the field work was during the winter and summer months. Quay listed 112 land birds, including 14 Falconiformes, 2 Columbiformes, 1 Cuculiformes, 5 Strigiformes, 2 Caprimulgiformes, 2 Apodiformes, 4 Piciformes, and 81 Passeriformes. He felt that the list was incomplete, particularly with reference to the passerine group during the spring and fall migration periods.

Davis (1958) and coworkers manned the first Operation Recovery station in North Carolina on 14 and 15 September 1957 at Caffeys Inlet on the Currituck Banks. Thirteen passerines of eight species were banded. Hailman and Hatch (1964) operated a banding station from 29 September to 2 October 1962, on Hatteras Island 3.2 km (2 miles) south of Salvo. They banded 61 individuals of 20 species in 88 net hours. During this same period Hailman and Hatch (1964) made sight observations during slack periods of mist-netting. Peacock (1964) intermittently operated banding stations at three localities from Duck to Kitty Hawk from 1956 through 1963. Each trip was of several days duration, sometimes in the late spring but mostly in the fall. During this 8-year period, 42 land-bird species were banded and several additional species were observed. Grey, Miller, Thompson, and Siler (1964) operated a banding station on Bodie Island from 24 through 26 September 1964. They banded a total of 72 birds of 25 species in 114 net hours. All of the individuals banded, except one, were land birds.

Although the frequency of brief notes in *The Chat* and *American Birds* on fall land-bird migration on the North Carolina barrier islands increased after 1960, most such studies prior to 1964 were short-term, erratic samplings. The need for systematic sampling of land birds throughout the fall migration period in this region prompted the initiation of my study in 1964 (Sykes 1967).

From 1955 through 1963, I made 14 one-day field trips to the Bodie Island-Pea Island area of Dare County, North Carolina, in late August and September to observe the avifauna. Between 24 August and 15 November 1964 I spent 18 days in the field on a preliminary survey of the autumn land-bird migration in this area. Data from these 32 field trips prior to 1965 indicated that large numbers of many land-bird species, many of which were out of their normal habitat and some not previously recorded there, occurred on these offshore barrier islands during the fall. Knowledge and experience gained from this earlier field work contributed much to the planning and execution of the principal research in 1965.

The purpose of my study was to conduct an extensive systematic investigation of all the land birds (Falconiformes, Columbiformes, Cuculiformes, Strigiformes, Caprimulgiformes, Apodiformes, Piciformes, and Passeriformes) occurring in the Bodie Island-Pea Island area during the fall migration period. This was the first such study along the southeastern coast of the United States south of Virginia. The specific objectives were to determine: (1) the species composition of the migration, (2) the magnitude, duration, and sequence of the migration, and (3) the general relationship of the migratory movements to the surface weather conditions.

METHODS

During the fall period, 21 August through 30 November 1965, intensive systematic censusing on established transects (= routes) was conducted daily, except when I was operating mist nets, when other business dictated absence from the study area, or when adverse weather (steady winds in excess of 15 knots, steady rain, or both) made field work impractical. Fifteen transects (Fig. 1) were designed to cover each type of terrestrial habitat and some marsh edge. Use of the available system of roads and trails made it possible for the entire study area to be surveyed adequately in a single day. The location of each transect was based on prior experience in the area. The complete transect route covered 66 km (41 miles); 18 km (11 miles) on foot and 48 km (30 miles) by automobile. Later in the fall the distance on foot was increased to facilitate more coverage on the ocean dunes and in the grassy habitats. This increase varied from 1.6 to 8 km (1-5 miles) per trip.

Field work began at dawn or shortly thereafter and ended at dusk. Half of the study area was covered in the morning and the other half in the afternoon, with a break of one to several hours during the middle of the day. I recorded the number of individuals of each species positively identified by sight and sound. In censusing, extreme care was taken to avoid, insofar as possible, counting the same individual twice on the same day. I used the techniques of "pishing," "swishing," and "squeaking" to attract secretive species from dense cover.

A banding station was operated on Bodie Island. This station was also a part of Operation Recovery for 1965. Mist nests were located at five sites in close proximity to each other. Up to 24 mist nets were used at a time.

The terms used in this publication for residence status are: (1) permanent — P, (2) transient — T, (3) summer — S, and (4) winter — W. In some species more than one residence status applies; these species are given two status codes. A permanent resident is a species that is found in the study area throughout the year. A transient is a bird that migrates through the area. A summer resident is a bird that nests in the area or nonbreeding individuals that occur during the nesting season. A winter resident is a bird that is found in the area from December through March.

Relative abundance as used in this publication refers only to the fall period. The terms used for relative abundance follow Stewart and Robbins (1958) and are: (1) abundant — A, means that a species, considering its habits and conspicuousness, was found in very large numbers; (2) common — C, means that a species, considering its habits and conspicuousness, was found in large numbers; (3) fairly common — F, means that a species, considering its habits and conspicuousness, was found in moderate or fair numbers; (4) uncommon — U, means that a species, considering its habits and conspicuousness, was found in rather small numbers; (5) rare — R, means that a species, within its normal range, was recorded in very small numbers; and (6) accidental — Ac, means that a species, well beyond its usual range, was recorded only once or twice.

A synopsis of field activity is presented in Tables 1 and 2. In Table 1 the type of activity is designated as follows: C — for censusing on transect, N — for mist-netting and banding, and NIF — for not in the field. The method by which the transect route was worked is given separately for the morning and the afternoon. SR, for standard route, is defined as a transect route in which all of the route normally worked on foot and by car in one calendar day was completed. NSR, for nonstandard route, is defined as a transect route in which some, but not all, of the route normally covered in one calendar day was completed. The transect route was covered by one of two procedures for both Bodie and Pea Islands. The procedures were as follows:

Bodie Island

B-1. Start in the vicinity of Seashore Maintenance Area, go north to Whalebone on N.C. Secondary Road 1243, then south on N.C. 12, covering Water Tower Road, Coquina Beach area, the road into the Visitors Center, Boneyard Road, Visitors Center, and finish in Pine Stand.

B-2. Start in Pine Stand, work the route in reverse of B-1, and finish in the vicinity of Seashore Maintenance Area.

Pea Island

P-1. Start at the north end of the island and work south, finishing at the refuge headquarters.

P-2. Start at the refuge headquarters and work the route in reverse of P-1.

Miscellaneous. In addition, the area around the marina at the southern tip of Bodie Island, when covered, was worked with the Pea Island section and is designated with an asterisk in the following manner — P-1* or P-2*. Under banding, Op indicates that the mist nets were in operation and NOp indicates that the mist nets were not in operation. P-NSR stands for Pea Island nonstandard route and B-NSR indicates the same for Bodie Island. Nets in Oper. means the amount of time the nets were open and in use. Net hours were computed by multiplying the number of nets in use by the number of hours in operation.

The daily weather data for this study were obtained from four main sources. While in the field, I recorded general weather conditions and readily observable changes throughout the day. Instrument readings were available from the U.S. National Park Service at the Bodie Island Ranger Station. This station collected weather information as a part of the U.S. Weather Bureau Cooperative Hurricane Reporting System, taking twice daily (8 a.m. and 4 p.m.) the sky condition, barometric pressure, wind direction and speed, and the amount of precipitation and the approximate or actual time that it occurred. The maximum and minimum temperatures for the preceding 24-hour period were recorded at 4 p.m. The broad spectrum of surface weather was obtained from the *Daily Weather Map* of the U.S. Department of Commerce, Weather Bureau (1965a). In addition, part of the criteria for determining cold-front passage and strength was obtained from *Local Climatological Data* for Cape Hatteras from the U.S. Department of Commerce, Weather Bureau (1965b).

Temperatures are in degrees Celsius (C) and wind speeds in knots. Clear sky is defined as the condition in which there is less than 25% cloud cover; partly cloudy, 25% to 75% cloud cover; and overcast, greater than 75% cloud cover. On days in which the 8 a.m. and 4 p.m. sky conditions differed, only the denser cloud condition was considered; thus, if 8 a.m.

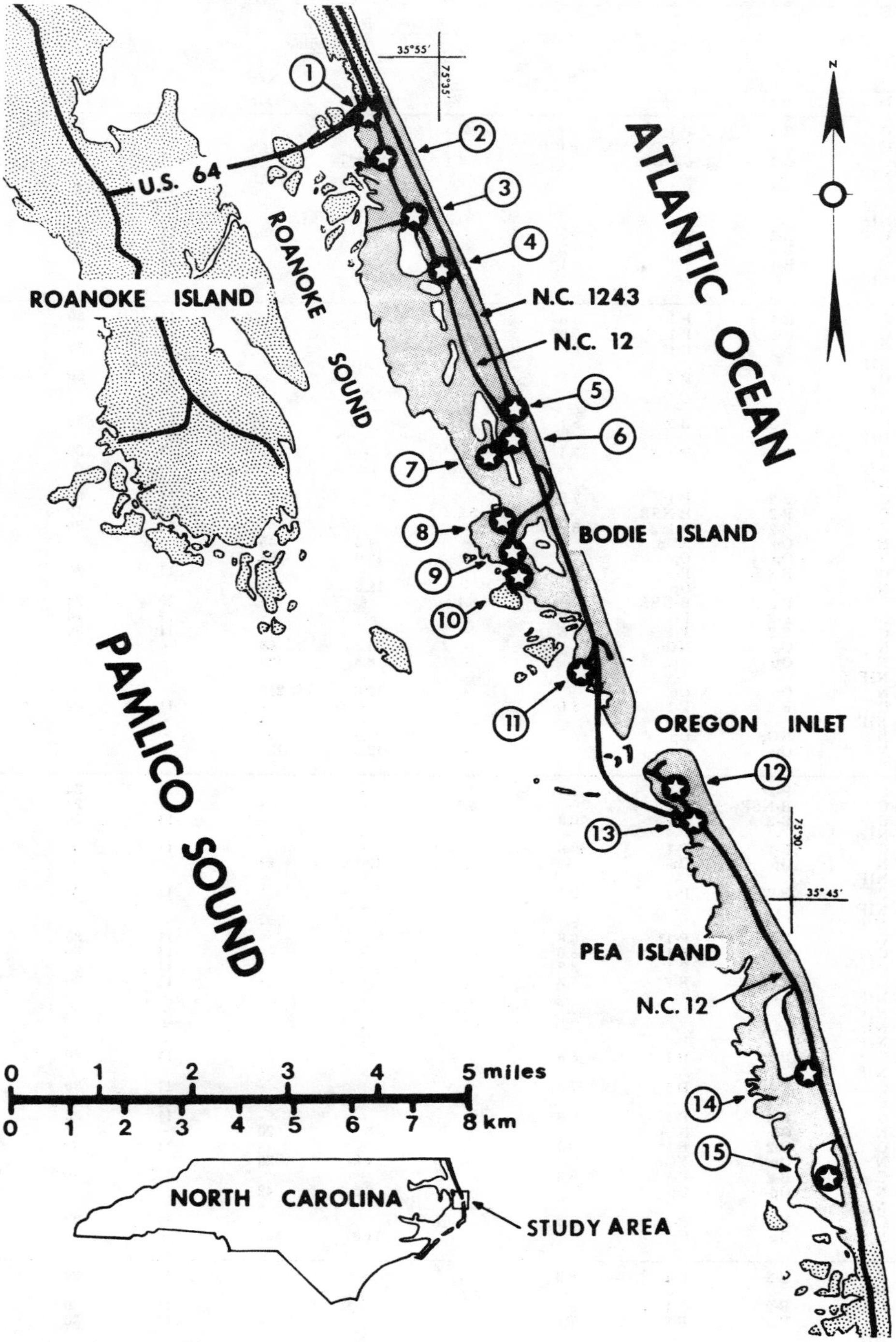

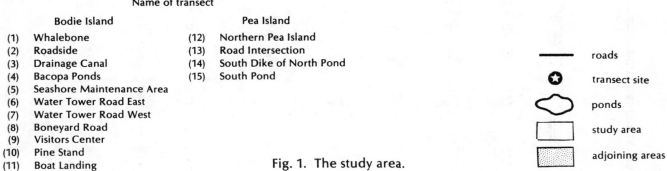

Name of transect

Bodie Island		Pea Island	
(1)	Whalebone	(12)	Northern Pea Island
(2)	Roadside	(13)	Road Intersection
(3)	Drainage Canal	(14)	South Dike of North Pond
(4)	Bacopa Ponds	(15)	South Pond
(5)	Seashore Maintenance Area		
(6)	Water Tower Road East		
(7)	Water Tower Road West		
(8)	Boneyard Road		
(9)	Visitors Center		
(10)	Pine Stand		
(11)	Boat Landing		

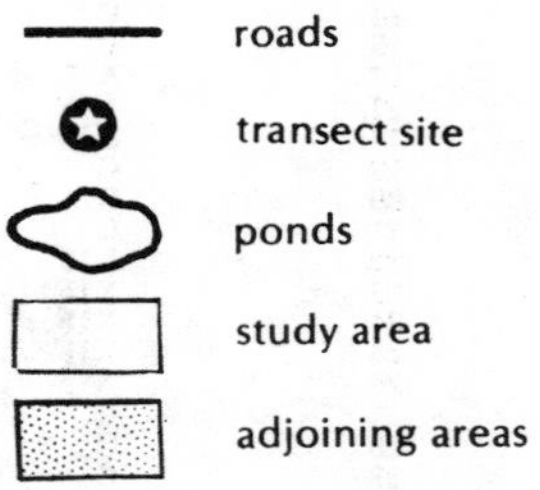

Fig. 1. The study area.

TABLE 1. Summary of daily field work, August through November 1965.[a]

					Hours				Mileage		Assistants	
					Censusing		Banding					
Date		Activity Type	a.m.	p.m.	SR	NSR	Nets in Oper.	Net Hours	On Foot	By Car	Censusing	Banding
Aug.	21	C	B-1	P-1	9.8	-	-	-	11	30	0	-
	22	C	B-1	P-1	8.0	-	-	-	11	30	0	-
	23	C	B-1	P-NSR	-	6.8	-	-	7	19	1	-
	24	NIF	-	-	-	-	-	-	-	-	-	-
	25	C	B-1	P-1	8.7	-	-	-	11	30	0	-
	26	NIF	-	-	-	-	-	-	-	-	-	-
	27	NIF	-	-	-	-	-	-	-	-	-	-
	28	C	B-1	P-1	9.5	-	-	-	11	30	2	-
	29	C	B-1	P-1	8.7	-	-	-	11	30	0	-
	30	C	B-1	P-1	9.4	-	-	-	11	30	0	-
	31	NIF	-	-	-	-	-	-	-	-	-	-
Sept.	1	C	B-1	P-1	8.8	-	-	-	11	30	0	-
	2	C	B-1	P-1	5.9	-	-	-	11	30	0	-
	3	C	B-1	P-1	8.2	-	-	-	11	30	0	-
	4	NIF	-	-	-	-	-	-	-	-	-	-
	5	C	B-1	P-1	7.5	-	-	-	11	30	0	-
	6	NIF	-	-	-	-	-	-	-	-	-	-
	7	NIF	-	-	-	-	-	-	-	-	-	-
	8	C	P-2	B-2	8.1	-	-	-	11	30	0	-
	9	NIF	-	-	-	-	-	-	-	-	-	-
	10	C	B-2	P-1	7.5	-	-	-	11	30	0	-
	11	NIF	-	-	-	-	-	-	-	-	-	-
	12	NIF	-	-	-	-	-	-	-	-	-	-
	13	C	B-2	P-1	7.8	-	-	-	11	30	0	-
	14	C	P-2	B-NSR	-	5.2	-	-	10	19	0	-
	15	C	B-2	P-1	6.9	-	-	-	11	30	0	-
	16	N	Op	NOp	-	-	4.5	45	-	-	-	0
	17	N	Op	Op	-	-	12.0	102	-	-	-	0
	18	C	B-2	P-1	9.6	-	-	-	11	30	2	-
	19	N	Op	Op	-	-	12.5	150	-	-	-	2
	20	C	P-2	B-NSR	-	8.1	-	-	10	24	0	-
	21	C	B-2	P-1	6.2	-	-	-	11	30	0	-
	22	C	B-2	P-1	6.8	-	-	-	11	30	0	-
	23	N	Op	NOp	-	-	5.5	66	-	-	-	1
	24	N	Op	Op	-	-	8.8	134	-	-	-	1
	25	NIF	-	-	-	-	-	-	-	-	-	-
	26	N	Op	Op	-	-	12.3	215	-	-	-	3
	27	C	B-2	P-1	8.6	-	-	-	11	30	0	-
	28	NIF	-	-	-	-	-	-	-	-	-	-
	29	N	NOp	Op	-	-	4.3	47	-	-	-	2
	30	N	Op	Op	-	-	12.5	138	-	-	-	2
Oct.	1	C	P-2*	B-2	7.0	-	-	-	11	24	0	-
	2	C	B-NSR	P-1	-	5.7	-	-	7	21	3	-
	3	C	B-2	P-1	10.3	-	-	-	11	30	3	-
	4	NIF	-	-	-	-	-	-	-	-	-	-
	5	C	B-2	P-1	10.8	-	-	-	11	30	1	-
	6	N	Op	Op	-	-	10.5	81	-	-	-	0
	7	NIF	-	-	-	-	-	-	-	-	-	-
	8	C	B-2	P-1	9.9	-	-	-	11	30	0	-
	9	NIF	-	-	-	-	-	-	-	-	-	-
	10	NIF	-	-	-	-	-	-	-	-	-	-
	11	C	B-2	P-1*	9.8	-	-	-	11	30	0	-
	12	C	B-2	P-1	8.6	-	-	-	11	30	0	-
	13	C	B-2	P-1	8.2	-	-	-	11	30	0	-
	14	C	P-2	B-2	7.3	-	-	-	11	30	0	-
	15	NIF	-	-	-	-	-	-	-	-	-	-
	16	NIF	-	-	-	-	-	-	-	-	-	-
	17	NIF	-	-	-	-	-	-	-	-	-	-
	18	NIF	-	-	-	-	-	-	-	-	-	-
	19	C	B-2	P-1	8.3	-	-	-	11	30	0	-
	20	NIF	-	-	-	-	-	-	-	-	-	-
	21	C	P-1	B-1	7.3	-	-	-	11	30	0	-
	22	C	B-2	-	-	4.8	-	-	5	15	0	-
	23	C	B-2	P-1	7.9	-	-	-	11	30	1	-
	24	N	Op	NOp	-	-	3.3	20	-	-	-	1
	25	C	B-2	P-1*	9.3	-	-	-	11	30	0	-
	26	N	Op	NOp	-	-	4.0	23	-	-	-	0
	27	C	B-2	P-1	9.3	-	-	-	11	30	1	-
	28	N	Op	Op	-	-	8.8	42	-	-	-	1
	29	NIF	-	-	-	-	-	-	-	-	-	-
	30	C	B-2	P-1	9.9	-	-	-	11	30	5	-
	31	N	Op	Op	-	-	11.8	161	-	-	-	5
Nov.	1	C	B-2	P-1	8.8	-	-	-	11	30	0	-
	2	NIF	-	-	-	-	-	-	-	-	-	-
	3	C	B-2	P-1*	9.1	-	-	-	11	30	0	-
	4	C	P-2*	B-2	7.4	-	-	-	11	30	0	-
	5	NIF	-	-	-	-	-	-	-	-	-	-
	6	C	B-2	P-1*	8.9	-	-	-	11	30	0	-
	7	NIF	-	-	-	-	-	-	-	-	-	-
	8	C	B-2	P-1*	8.7	-	-	-	11	30	0	-
	9	N	Op	Op	-	-	7.8	93	-	-	-	1
	10	NIF	-	-	-	-	-	-	-	-	-	-
	11	C	B-2	P-1*	6.8	-	-	-	12	30	0	-
	12	NIF	-	-	-	-	-	-	-	-	-	-
	13	NIF	-	-	-	-	-	-	-	-	-	-
	14	C	B-2	P-1	8.3	-	-	-	12	30	0	-
	15	C	P-1	B-2	7.8	-	-	-	12	30	0	-
	16	NIF	-	-	-	-	-	-	-	-	-	-
	17	NIF	-	-	-	-	-	-	-	-	-	-
	18	NIF	-	-	-	-	-	-	-	-	-	-
	19	C	B-2	P-1	7.8	-	-	-	12	30	0	-
	20	C	B-2	P-1	7.1	-	-	-	12	30	0	-
	21	C	B-NSR	P-NSR	-	8.9	-	-	20	0	0	-
	22	NIF	-	-	-	-	-	-	-	-	-	-
	23	NIF	-	-	-	-	-	-	-	-	-	-
	24	NIF	-	-	-	-	-	-	-	-	-	-
	25	NIF	-	-	-	-	-	-	-	-	-	-
	26	NIF	-	-	-	-	-	-	-	-	-	-
	27	C	B-2	P-1	7.5	-	-	-	12	30	0	-
	28	C	B-2	P-1	8.1	-	-	-	12	30	0	-
	29	C	P-1	B-2	8.3	-	-	-	15	30	0	-
	30	C	P-1	B-2	8.8	-	-	-	16	22	0	-

[a]C — censusing, N — mist-netting and banding, NIF — not in the field,
B — Bodie Island, P — Pea Island, SR — standard census route, NSR — nonstandard census route,
OP — mist nets in operation, NOp — mist nets not in operation; see text for further explanation.

TABLE 2. Summary of total field work for 102 days, 21 August through 30 November 1965.[a]

	August	September	October	November	Total
Activity (in days)					
Censusing	7	14	16	15	52
Banding	0	8	5	1	14
Total in the field	7	22	21	16	66
Not in the field	4	8	10	14	36
Percent of days in the field	64	73	68	53	65
Hours					
Censusing	60.9	105.2	134.4	122.3	422.8
Standard Route	54.1	91.9	123.9	113.4	383.3
Nonstandard Route	6.8	13.3	10.5	8.9	39.5
Banding					
Nets in operation	0.0	72.4	38.4	7.8	118.6
Net hours	0	897	327	93	1317
Total in the field	60.9	177.6	172.8	130.1	541.4
Average number/day censusing	8.7	7.5	8.4	8.2	8.1
Average number/day banding	0.0	9.1	7.7	7.8	8.5
Average number/day in the field	8.7	8.1	8.2	8.1	8.2
Mileage					
On foot	73	152	166	190	581
By car	199	403	450	412	1464
Total	272	555	616	602	2045
Average/day on foot	10.4	10.9	10.4	12.7	11.2
Average/day by car	28.4	28.8	28.1	27.5	28.2
Average/day total	38.9	39.6	38.5	40.1	39.3
Number of Assistants					
Censusing	3	2	14	0	19(13)
Banding	0	11	7	1	19(13)
Total	3(3)	13(8)	21(10)	1(1)	38(18)

[a]Figures in parentheses designate number of different persons.

was clear and 4 p.m. was overcast, the day was classed as overcast.

The scientific names of plants used in this publication follow Radford et al. (1964), and the common names of plants are the author's preference. The scientific names, common names, and phylogenetic sequence of birds follow the *Check-list of North American Birds*, sixth edition (AOU 1983). A complete list of the scientific names of birds recorded during the study is given in Appendix B.

STUDY AREA

The study area (Fig. 1) was that sector of the barrier islands (beaches) of Dare County, N.C., between Whalebone and the present headquarters of Pea Island National Wildlife Refuge. These barrier beaches consist of a peninsula and a chain of low-lying islands that extend from southeastern Virginia southward along the coast of North Carolina to Cape Fear. That portion of the barrier islands between Oregon Inlet and Ocracoke Inlet is known as the Outer Banks[1]. Bodie Island lies to the north of Oregon Inlet and Pea Island to the south. The mainland ranges from 15 to 19 km (9-12 miles) to the west across Pamlico Sound. Roanoke Island is 1.9 km (1.2 miles) across Roanoke Sound to the west of Bodie

[1] In recent years the term "Outer Banks" has become widely accepted in reference to the entire North Carolina coast from the Virginia line south to Ocracoke and occasionally even to the Cape Lookout area.

Island and lies between the barrier beach and the mainland. To the east of Bodie and Pea Islands is the Atlantic Ocean. Bodie Island is no longer an island, but the southern tip of a long, narrow peninsula that is connected to the mainland in Virginia, owing to the closing of Roanoke, Caffeys (formerly called Carthys), and Currituck Inlets in the early 1800s (Dunbar 1958). In 1945 New Inlet closed (Dunbar 1958), joining Pea Island with Hatteras Island. Even though Bodie Island and Pea Island are both physically connected to other much larger land masses, the two names are retained for their respective areas.

None of the small islands lying along the western shore and at Oregon Inlet were included in the study. Bodie Island is within the Cape Hatteras National Seashore Recreation Area, except for a narrow strip in private ownership. Situated along the ocean from Whalebone south to the Seashore Maintenance Area, the private property has its western boundary about midway between N.C. 12 and N.C. 1243. All of the study area on Pea Island lies within the refuge, which itself is geographically within the Cape Hatteras National Seashore.

The study area is 26.9 km (16.7 miles) in length, ranges from 0.8 to 2.4 km (0.5-1.5 miles) in width, and lies along a NNW axis. Total area for the terrestrial and marsh environs is approximately 3,266 ha (8,070 acres); 2,048 ha (5,060 acres) on Bodie Island and 1,218 ha (3,010 acres) on Pea Island. Sizes of the areas were determined from U.S. Geological Survey Quadrangles of the 7.5-minute Series (Topographic) with a scale of 1:24000 by use of a compensating polar planimeter. The acreage given excludes all large ponds, creeks, and the adjacent small islands in the sounds and in Oregon Inlet.

Topographic variations are relatively slight. The islands are basically flat strands of sand with dunes along the open beach and marshes toward the sound. Elevations range from zero to 9.1 m (30 feet) above mean sea level. However, the mean elevation is less than 1.5 m (5 feet). The narrow dune line behind the open beach attains a height of 9.1 m (30 feet) in places. On the sound at the site of the water tower on Bodie Island, isolated dunes, surrounded by marsh on three sides, reach an elevation of 7.3 m (24 feet). Ponds are numerous; to the west of N.C. 12 on Bodie Island there are seven large ponds, and on Pea Island there is a small pond at the north end in addition to three large ponds to the west of N.C. 12 between the U.S. Coast Guard Station and the refuge headquarters. During heavy rains many small intermittent pools form in the low-lying areas and depressions. The three large ponds on Pea Island are, from north to south, North Pond, New Field Pond, and South Pond. These ponds are surrounded on three sides by a dike system with the dunes along the ocean front and N.C. 12 forming the fourth side. At the time of this study, New Field Pond did not exist, as the dike that created this pond was not built until the late 1970s. The ponds on Pea Island are fresh, whereas on Bodie Island they are brackish or saline. The shoreline on the ocean is relatively smooth in contrast to the rather irregular shoreline on the sound, which is broken by many small creeks and bays. In the sounds, close inshore, are numerous small, marshy islands. Several islands have also been created in Pamlico Sound to the west of Oregon Inlet as the result of dredging activity.

The substrate is basically sand and shell. In certain areas, principally the brackish and fresh marshes and shrub thickets on low moist sites, there is a substantial accumulation of organic material. On the drier sites, however, there is little or no such accumulation. Soils are in a very early stage of development (i.e. the stratification that is characteristic of a mature soil is poorly defined or lacking).

Major Habitats

The terrestrial habitats of the study area are classified under eight major types, most of which closely follow the work of Quay (1959). The marine, brackish, and fresh bodies of water are not discussed as habitats *per se*. The major habitats are: (1) beach, (2) dune, (3) herb-shrub, (4) shrub thicket, (5) pine thicket, (6) fresh marsh, (7) high marsh, and (8) low marsh.

In an idealized transect across a barrier island (Fig. 2), from the ocean west to the sound, the habitats lie in parallel zones in sequence as listed above. In actuality only the beach, dune, herb-shrub, and high marsh are found in almost continuous, narrow, parallel zones. Widespread intermingling of habitats results in extensive ecotones.

Beach. This zone extends from the mean

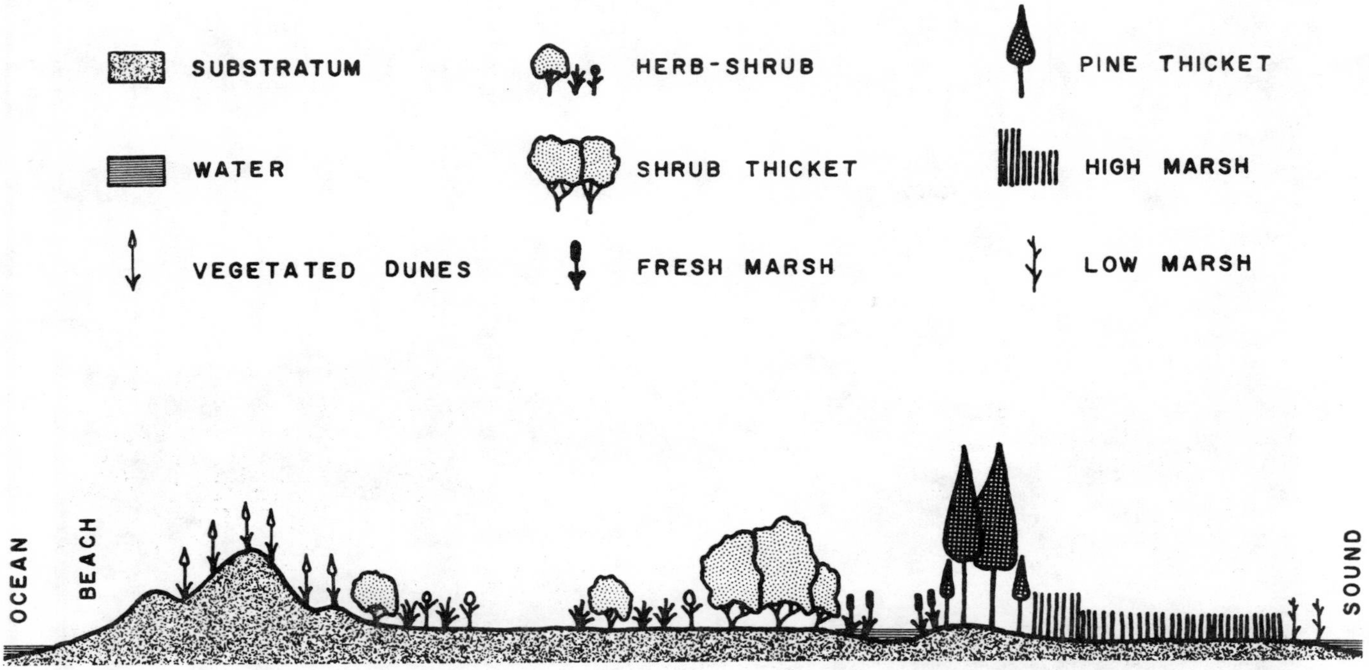

Fig. 2. An idealized transect across the study area, from ocean to sound.

low water mark to the base of the dunes. In the uppermost portions of the beach, above the high-tide line, there may be some sparse vegetation consisting mainly of seedling Sea Oats (*Uniola paniculata*) and Sea Rocket (*Cakile* spp.). During storms or extremely high tides, the zone is completely inundated by salt water.

Dune. The ocean dunes (Fig. 3) form a narrow zone that is more or less parallel with the shoreline. This zone is subject to heavy salt spray, much wind erosion, abrasion by wind-blown sand, high ground temperatures during the growing season, and dry conditions. On Bodie Island there is an area of isolated dunes toward the sound. Most of the dunes are fully vegetated. In recent years the elimination of grazing, protection from fire, widespread planting of grasses, and the application of fertilizer greatly facilitated the restoration and development of plant growth on the dunes, resulting in their stabilization. The dominant plants of this zone are: Sea Oats, Beachgrass (*Ammophila breviligulata*), Slat Meadow Grass (*Spartina patens*), and, on Pea Island only, Wild Bean (*Strophostyles helvola*).

Herb-shrub. To the leeward side of the ocean dunes is the herb-shrub zone (Fig. 4, 5), which covers extensive areas and intermingles with the dunes to the east and the shrub thickets and fresh and high marshes to the west. This zone is characterized by isolated woody plants dispersed throughout areas of grasses and forbs. Frequently found within the zone are small scattered patches of bare sand. The shoulders of the roads and grassy areas around buildings are somewhat atypical, in that they are mowed regularly, thus eliminating the shrub component. The shoulders of the roads are mowed for a distance of 3 to 7.6 m (10-25 feet) from the pavement or gravel surface, and the grassy areas around buildings are mowed for a distance of 7.6 m (25 feet) to more than 30.5 m (100 feet) from these structures. The major plant species of the herb-shrub zone are: Annual Bluegrass (*Poa annua*), Salt Meadow Grass, Bermuda Grass (*Cynodon dactylon*), Crab Grass (*Digitaria sanguinalis*), Knotgrass (*Paspalum* spp.), Panic Grass (*Panicum* spp.), Sandbur (*Cenchrus tribuloides*), Broomsedge (*Andropogon virginicus*), Bayberry (*Myrica pensylvanica*), Wax Myrtle (*Myrica cerifera*), Live Oak (*Quercus virginiana*), Poke (*Phytolacca americana*), Black Cherry (*Prunus serotina*), Wild Bean, Yaupon (*Ilex vomitoria*), Marsh Elder (*Iva frutescens*), Groundsel (*Baccharis halimifolia*), and Seaside Goldenrod (*Solidago sempervirens*).

Fig. 3. The leeward side of vegetated ocean dunes on Pea Island.

Shrub Thicket. The shrub-thicket zone (Fig. 6) occurs in the transition from the upland sites into the high marsh. This habitat type is typically in broken patches intermingled with the two major adjoining zones. The major species of this zone are: Loblolly Pine (*Pinus taeda*), Pond Pine (*Pinus serotina*), Italian Cluster Pine (*Pinus pinaster*) [planted in several locations over sizable acreage in the 1930s on Bodie Island], Red Cedar (*Juniperus virginiana*), Greenbriar (*Smilax* spp.), Bayberry, Wax Myrtle, Live Oak, Coastal Plain Willow (*Salix caroliniana*), Red Bay (*Persea borbonia*), Blackberry (*Rubus* spp.), Black Cherry, Poison Ivy (*Rhus radicans*), Yaupon, Muscadine Grape (*Vitis rotundifolia*), and Groundsel.

Pine Thicket. The pine-thicket habitat (Fig. 7) occurs in small, isolated stands in several places on Bodie Island and is altogether absent on Pea Island. The trees in such stands range in height from 1.8 to 16.8 m (6-55 feet). Typically, the crowns of the trees form a complete dome-shaped canopy. The dominant species in this type are Loblolly Pine and Pond Pine.

Fresh Marsh. The fresh-marsh habitat is found only around the edge of the three ponds on Pea Island. The dominant species are: Cattail (*Typha* spp.), Salt Meadow Grass, Spike Rush (*Eleocharis* spp.), Three-square (*Scirpus americanus*), Black Needlerush (*Juncus roemerianus*), Rush (*Juncus scirpoides*), Smartweed (*Polygonum* spp.), and Marsh Mallow (*Hibiscus moscheutos*).

High Marsh. The high marsh (Fig. 8) occupies a nontidal zone between the upland communities and the shore of the sound, except where it is replaced by low marsh in the vicinity of the inlet and in scattered locations on Pea Island. Conditions in this habitat are basically brackish. The marsh is not usually flooded by sound or sea water except during strong winds or storms. The numerous ponds on Bodie Island lie within this zone. This habitat is characterized by two distinct vegetation types, almost pure stands of Black Needlerush and extensive stands of Salt

Fig. 4. Typical herb-shrub habitat.

Fig. 5. Herb-shrub habitat, showing the major components—areas of bare sand, grasses and forbs, and shrubs.

Meadow Grass. This latter type also contains some Spike Grass (*Distichlis spicata*), Bulrush (*Scirpus robustus*), and a few other species. The major species of the zone are Spike Grass, Salt Meadow Grass, Salt Reed Grass (*Spartina cynosuroides*), Spike Rush, Sand Rush (*Fimbristylis castanea*), Bulrush, Black Needlerush, Seashore Mallow (*Kosteletskya virginica*), Marsh Mallow, Marsh Elder, Groundsel, and Sea Ox-eye (*Borrichia frutescens*).

Low Marsh. The low marsh is a tidal marsh. On Bodie Island it covers a small area along the shore of the sound in the vicinity of the inlet, while on Pea Island it is widespread from the vicinity of the inlet southward, in discontinuous stands, to the southern boundary of the study area. The major species of the low marsh are Saltwater Cordgrass (*Spartina alterniflora*) and Black Needlerush.

Fig. 6. A dense shrub thicket containing small trees.

Fig. 7. Mature pine thicket south of the visitors center on Bodie Island.

Climate

The study area has a maritime climate, with moderate temperatures throughout the year. The average annual temperature is about 17°C, with temperatures as low as 0°C on an average of 14 days a year. On only one day in an average year does the temperature fail to rise above the freezing point. In summer, temperatures as high as 32°C occur on an average of only once annually. The rainfall is usually fairly well distributed throughout the year, with an annual average of 127+ cm (50 inches); however, the greatest amounts occur in July and August. The prevailing winds are from the SW in spring and summer and from the NE and N in the fall and winter. The mean hourly wind speed is 11 knots, compared with 9 knots on the coast of the mainland and 6 to 7 knots inland in North Carolina (Carney and Hardy 1964).

Fig. 8. High marsh, with shrub thicket in the background.

RESULTS

Magnitude and Duration of the Migration

A grand total of 109,192 individuals of 145 species (Table 3) were recorded on the census transects: 7,587 individuals of 69 species in August; 15,416 of 102 species in September; 65,629 of 120 species in October; and 20,560 of 76 species in November. An average of 48 land-bird species were recorded on each census day, excluding the six days on which a nonstandard transect route was worked. Birds that were banded or collected are not included in Table 3.

On the censuses in 1965, 53.8% of the birds were transients, 17.2% winter residents and transients, 11.7% permanent residents and transients, 6.9% summer residents and transients, 5.5% permanent residents, 3.5% accidentals, and 1.4% summer residents. Relative abundance in relation to residence status is summarized in Table 4. Of the 145 species recorded, 56% were classed as uncommon, rare, or accidental, while only 44% were considered to be abundant, common, or fairly common. Fairly common transients accounted for 12% of the species, while uncommon transients were 16.6% and rare transients 22%.

A total of 1,276 individuals of 60 species were banded (Table 5) on 14 days of mist-netting, with a mean of 0.97 birds per net hour. The mean number of species per day of netting was 14.6 and the mean number of individuals captured per day was 91. The most common species banded (total of 15 or more individuals in descending order for fall) were Yellow-rumped Warbler, American Redstart, Gray Catbird, Common Yellowthroat, Brown Creeper, Dark-eyed Junco, Swainson's Thrush, Blackpoll Warbler, Magnolia Warbler, Cape May Warbler, Swamp Sparrow, Black-throated Blue Warbler, and Golden-crowned Kinglet. No species were captured in the nets that were not also recorded on the censuses, and no birds were captured that had been banded at another locality (foreign retrap).

Eleven individuals of 10 species were collected (Table 6) during the 102-day period. In

TABLE 3. List of birds recorded on the 52 censuses, August through November 1965[a]

Species	% Days Recorded	Total Individuals[b] Aug.	Sept.	Oct.	Nov.	Fall	Occurrence[c]	Peaks Date	No.	Residence Status[d]	Relative Abundance[e]
Turkey Vulture	2	0	0	1	0	1	10/27	-	-	P	R
Osprey	67	3	15	31	12	61	8/23-11/8	-	-	S,T	U
Bald Eagle	2	0	0	1	0	1	10/5	-	-	P	R
Northern Harrier	87	4	28	95	117	244	8/28-11/30	-	-	W,T	C
Sharp-shinned Hawk	46	0	2	40	25	67	9/27-11/30	10/23	16	T	F
Cooper's Hawk	2	0	0	1	0	1	10/21	-	-	T	R
Red-shouldered Hawk	54	3	9	11	10	33	8/21-11/29	-	-	P	U
Swainson's Hawk	2	0	0	1	0	1	10/11	-	-	-	Ac
Red-tailed Hawk	4	0	0	1	1	2	10/30-11/8	-	-	T	R
American Kestrel	81	0	167	155	137	459	9/1-11/30	9/13	42	W,T	C
								10/3	22		
Merlin	63	0	6	43	22	71	9/13-11/30	10/21	7	W,T	F
								10/30	7		
Peregrine Falcon	38	0	2	14	8	24	9/5-11/29	-	-	W,T	U
Mourning Dove	94	178	448	508	66	1,200	8/21-11/30	9/20	60	P,T	C
								10/13	101		
								10/25	33		
Black-billed Cuckoo	13	4	3	1	0	8	8/22-10/5	-	-	T	U
Yellow-billed Cuckoo	38	6	3	23	1	33	8/21-11/1	-	-	S,T	U
Common Barn-Owl	25	2	2	3	11	18	8/21-11/15	-	-	W,T	U
Short-eared Owl	6	0	0	0	3	3	11/3-11/30	-	-	W,T	U
Common Nighthawk	8	2	0	2	0	4	8/21-10/23	-	-	S	R
Chimney Swift	25	9	7	4	0	20	8/21-10/13	-	-	S	U
Ruby-throated Hummingbird	12	15	3	0	0	18	8/21-9/1	8/25	8	T	U
Red-headed Woodpecker	13	1	2	7	0	10	8/30-10/27	-	-	T	U
Yellow-bellied Sapsucker	12	0	3	38	0	41	9/27-10/11	10/5	18	T	F
Downy Woodpecker	92	19	39	83	51	192	8/21-11/30	-	-	P,T	F
Hairy Woodpecker	31	0	0	6	23	29	10/23-10/28	-	-	T	U
Northern Flicker	100	17	167	847	460	1,491	8/21-11/30	9/15	22	P,T	A
								9/27	44		
								10/5	122		

Olive-sided Flycatcher	4	0	3	0	0	3	9/13-9/18	-	-	T	R
Eastern Wood-Pewee	25	2	10	13	0	25	8/29-10/12	-	-	T	U
Yellow-bellied Flycatcher	10	4	1	0	0	5	8/21-9/1	-	-	T	U
Acadian Flycatcher	2	1	0	0	0	1	8/21	-	-	T	R
Empidonax species	33	12	27	4	0	43	8/25-10/8	8/30	7	T	U
Eastern Phoebe	33	0	1	52	4	57	9/27-11/8	10/5	20	T	F
Great Crested Flycatcher	35	25	33	3	0	61	8/21-10/1	8/30	8	T	F
Western Kingbird	17	0	3	5	2	10	9/10-11/15	-	-	T	U
Eastern Kingbird	40	252	345	2	0	599	8/21-10/3	8/21	39	S,T	C
								8/30	73		
								9/3	32		
								9/13	50		
Purple Martin	21	93	9	0	0	102	8/21-9/13	8/22	31	S,T	C
Tree Swallow	25	5	1,333	43,742	387	45,467	8/21-11/30	9/20	510	W,T	A
								10/3	3,300		
								10/12	16,000		
								10/19	8,300		
								11/1	185		
N. Rough-winged Swallow	2	0	1	0	0	1	9/1	-	-	T	R
Bank Swallow	6	82	2	0	0	84	8/21-9/18	8/28	81	T	U
Cliff Swallow	4	0	1	2	0	3	9/18-10/12	-	-	T	R
Barn Swallow	48	1,066	271	41	0	1,378	8/21-10/14	8/28	938	S,T	A
								9/8	180		
American Crow	58	5	12	20	17	54	8/22-11/30	-	-	P	U
Fish Crow	77	566	885	284	34	1,769	8/21-11/27	9/5	105	P,T	C
Carolina Chickadee	77	19	42	19	14	94	8/21-11/29	-	-	P,T	U
Red-breasted Nuthatch	62	10	20	106	6	142	8/29-11/2	8/30	6	T	F
								10/5	22		
								10/25	16		
White-breasted Nuthatch	4	0	0	3	0	3	10/5-10/19	-	-	T	R
Brown Creeper	40	0	4	167	7	178	9/27-11/19	10/5	34	T	C
								10/11	14		
								10/25	40		
Carolina Wren	98	171	235	240	159	805	8/21-11/30	-	-	P	C
House Wren	62	0	6	97	80	183	9/15-11/29	10/3	12	T	F
								10/22	11		
								11/3	10		
								11/6	11		
								11/19	11		
Winter Wren	4	0	0	4	0	4	10/8-10/27	-	-	T	U
Sedge Wren	44	2	2	12	43	59	8/29-11/30	-	-	W,T	C
Marsh Wren	38	2	2	12	23	39	8/22-11/29	-	-	P,T	F

Table 3. (continued)

Species	% Days Recorded	Total Individuals[b] Aug.	Sept.	Oct.	Nov.	Fall	Occurrence[c]	Peaks Date	No.	Residence Status[d]	Relative Abundance[e]
Golden-crowned Kinglet	31	0	0	153	29	182	10/5-11/30	10/25	75	W,T	C
Ruby-crowned Kinglet	38	0	1	221	64	286	9/27-11/30	10/5	19	W,T	C
								10/25	81		
								10/30	65		
Blue-gray Gnatcatcher	8	2	1	2	0	5	8/22-10/14	-	-	T	U
Veery	2	0	1	0	0	1	8/5	-	-	T	R
Gray-cheeked Thrush	8	1	1	5	0	7	8/30-10/5	10/5	4	T	U
Swainson's Thrush	17	0	11	26	0	37	9/10-10/14	10/5	15	T	F
Hermit Thrush	25	0	0	29	16	45	10/13-11/19	-	-	T	F
Wood Thrush	2	0	0	1	0	1	10/5	-	-	T	R
American Robin	56	0	1	264	295	560	9/27-11/30	10/25	68	W,T	F
								11/27	56		
Gray Catbird	98	490	732	552	246	2,020	8/21-11/30	8/23	106	P,T	A
								8/30	72		
								9/14	67		
								10/3	53		
								10/11	63		
								11/19	35		
Northern Mockingbird	98	148	428	428	317	1,321	8/21-11/30	9/15	55	P,T	A
								10/13	50		
Sage Thrasher	2	0	0	1	0	1	10/5	-	-	-	Ac
Brown Thrasher	92	68	303	197	38	606	8/21-11/29	9/14	36	P,T	C
								9/18	42		
								10/3	36		
								10/30	12		
Water Pipit	17	0	0	2	88	90	10/2-11/30	-	-	W,T	U
Sprague's Pipit	2	0	0	0	1	1	11/30	-	-	-	Ac
Cedar Waxwing	60	12	41	113	30	196	8/25-11/28	10/12	27	T	F
Loggerhead Shrike	10	2	3	0	0	5	8/21-9/8	-	-	T	R
European Starling	100	281	522	1,352	1,839	3,994	8/21-11/30	8/25	73	P,T	C
								9/8	87		
								10/3	120		
								10/12	60		
								10/23	213		
								10/27	238		
								11/3	216		
								11/14	180		

White-eyed Vireo	38	24	38	2	0	64	8/21-10/12	9/10	8	S,T	F
Solitary Vireo	4	0	0	2	0	2	10/5-10/11	-	-	T	R
Philadelphia Vireo	2	0	0	1	0	1	10/2	-	-	T	R
Red-eyed Vireo	10	0	8	7	0	15	9/8-10/5	-	-	T	U
Blue-winged Warbler	2	0	1	0	0	1	9/18	-	-	T	R
Tennessee Warbler	10	1	2	6	0	9	8/30-10/8	10/5	5	T	U
Orange-crowned Warbler	8	0	0	5	0	5	10/1-10/23	-	-	W,T	U
Nashville Warbler	12	1	3	3	0	7	8/29-10/25	-	-	T	U
Northern Parula	15	0	6	10	0	16	9/18-10/13	-	-	T	U
Yellow Warbler	27	20	15	0	0	35	8/21-9/27	-	-	S,T	F
Chestnut-sided Warbler	4	0	1	1	0	2	8/18-10/5	-	-	T	R
Magnolia Warbler	21	2	12	16	0	30	8/30-10/25	10/5	10	T	F
Cape May Warbler	38	1	31	83	0	115	8/22-10/25	9/8	10	T	C
								9/27	12		
								10/5	23		
								10/11	15		
Black-thr. Blue Warbler	19	0	10	15	0	25	9/8-10/27	10/5	8	T	F
Yellow-rumped Warbler	67	0	22	4,215	7,582	11,819	9/13-11/30	9/27	19	W,T	A
								10/11	400		
								10/25	875		
								11/8	978		
								11/19	915		
								11/28	685		
Black-thr. Green Warbler	12	0	2	7	0	9	9/27-10/30	-	-	T	U
Blackburnian Warbler	6	0	1	2	0	3	9/27-10/11	-	-	T	R
Yellow-throated Warbler	6	1	0	2	0	3	8/21-10/8	-	-	T	R
Pine Warbler	10	0	1	3	3	7	9/27-11/19	-	-	T	U
Prairie Warbler	50	56	78	11	0	145	8/21-10/30	8/25	14	S,T	C
								9/8	14		
								9/18	16		
Palm Warbler	67	0	54	634	38	726	9/8-11/30	9/27	40	W,T	A
								10/3	63		
								10/14	121		
								10/23	44		
Bay-breasted Warbler	6	0	1	3	0	4	9/8-10/5	-	-	T	R
Blackpoll Warbler	23	0	5	64	0	69	9/27-10/25	10/5	18	T	F
Black-and-white Warbler	37	13	22	22	0	57	8/21-10/8	10/3	8	T	F
								10/5	8		
American Redstart	65	36	223	159	0	418	8/21-10/25	9/8	28	T	A
								9/18	34		
								9/27	70		
								10/3	50		
Prothonotary Warbler	8	4	1	0	0	5	8/21-9/8	-	-	T	U
Ovenbird	2	0	2	0	0	2	9/27	-	-	T	R
Northern Waterthrush	37	13	15	6	1	35	8/21-11/1	-	-	T	F
Connecticut Warbler	4	0	2	0	0	2	9/1-9/8	-	-	T	R

Table 3. (continued)

Species	% Days Recorded	Total Individuals[b] Aug.	Sept.	Oct.	Nov.	Fall	Occurrence[c]	Peaks Date	No.	Residence Status[d]	Relative Abundance[e]
Common Yellowthroat	98	328	654	324	39	1,345	8/21-11/29	9/8	89	P,T	A
								9/18	98		
								10/3	56		
Hooded Warbler	2	1	0	0	0	1	8/22	-	-	T	R
Wilson's Warbler	2	0	0	2	0	2	10/5	-	-	T	R
Canada Warbler	4	2	0	0	0	2	8/21-8/29	-	-	T	R
Yellow-breasted Chat	13	4	5	0	0	9	8/22-9/18	-	-	T	U
Summer Tanager	6	0	1	1	0	2	9/8-10/13	-	-	T	R
Scarlet Tanager	4	0	2	4	0	6	9/27-10/21	-	-	T	R
Northern Cardinal	88	45	58	46	41	190	8/21-11/30	-	-	P	F
Rose-breasted Grosbeak	8	0	2	4	0	6	9/20-10/5	-	-	T	U
Blue Grosbeak	15	0	3	7	1	11	9/18-11/3	-	-	T	U
Indigo Bunting	27	0	11	23	0	34	9/1-10/25	10/5	8	T	U
Dickcissel	6	1	0	2	0	3	8/25-10/3	-	-	T	R
Rufous-sided Towhee	98	320	342	357	311	1,330	8/21-11/30	-	-	P	A
Bachman's Sparrow	2	0	0	1	0	1	10/11	-	-	T	R
American Tree Sparrow	6	0	0	3	3	6	10/30-11/3	-	-	T	R
Chipping Sparrow	44	0	3	57	38	98	9/18-11/29	10/8	12	T	F
								11/1	19		
Clay-colored Sparrow	12	0	0	7	2	9	10/5-11/14	-	-	T	U
Field Sparrow	96	82	152	132	65	431	8/21-11/30	9/14	23	P,T	C
								10/3	23		
								11/1	20		
Vesper Sparrow	21	0	0	26	11	37	10/19-11/11	-	-	T	F
Lark Sparrow	44	18	30	13	2	63	8/21-11/20	9/27	6	T	F
Savannah Sparrow	73	2	40	1,178	518	1,738	8/21-11/30	10/5	108	W,T	A
								10/11	191		
								10/25	118		
(Ipswich race)	2	0	0	0	1	1	11/27	-	-	W,T	R
Grasshopper Sparrow	21	0	1	8	5	14	9/18-11/28	-	-	T	U
Sharp-tailed Sparrow	8	0	0	10	2	12	10/11-11/1	10/30	8	W,T	F
Seaside Sparrow	8	2	1	2	0	5	8/25-10/25	-	-	P,T	F
Fox Sparrow	10	0	0	2	7	9	10/30-11/19	-	-	W,T	U
Song Sparrow	100	124	213	407	379	1,123	8/21-11/30	-	-	P,T	C
Lincoln's Sparrow	12	0	0	6	1	7	10/5-11/11	-	-	T	U
Swamp Sparrow	46	0	0	69	406	475	10/5-11/30	11/19	51	W,T	C
White-throated Sparrow	54	0	4	289	84	377	9/14-11/29	10/5	14	W,T	C
								10/30	85		

White-crowned Sparrow	40	0	0	134	70	204	10/3-11/20	10/30	42	T	F
								11/3	26		
Dark-eyed Junco	58	0	6	1,296	897	2,199	9/27-11/30	10/19	77	W,T	A
								10/27	388		
								11/1	391		
Snow Bunting	4	0	0	0	25	25	11/21-11/29	-	-	W,T	U
Bobolink	44	15	150	27	0	192	8/28-10/22	9/15	61	T	F
Red-winged Blackbird	100	2,036	5,339	3,422	2,910	13,707	8/21-11/30	-	-	P,T	A
Eastern Meadowlark	100	66	165	420	600	1,251	8/21-11/30	-	-	P,T	A
Western Meadowlark	2	0	0	0	1	1	11/30	-	-	-	Ac
Rusty Blackbird	12	0	0	5	2	7	10/3-11/28	-	-	T	R
Brewer's Blackbird	2	0	0	0	1	1	11/14	-	-	-	Ac
Boat-tailed Grackle	100	326	1,011	1,587	1,658	4,582	8/21-11/30	9/21	207	P,T	A
								10/13	237		
Common Grackle	4	2	0	0	0	2	8/22-8/23	-	-	S,T	R
Brown-headed Cowbird	58	78	185	217	1	481	8/21-11/3	8/21	27	S,T	F
								9/3	34		
								9/22	63		
								10/3	61		
								10/14	25		
								10/22	24		
Northern Oriole	54	340	214	45	0	599	8/21-10/30	8/30	218	T	A
								9/8	50		
								9/18	29		
								10/3	24		
Purple Finch	12	0	1	6	1	8	10/27-11/28	-	-	W,T	R
Common Redpoll	2	0	0	0	1	1	11/11	-	-	T	R
Pine Siskin	31	0	0	10	36	46	10/5-11/30	-	-	W,T	U
American Goldfinch	21	0	0	0	67	67	11/6-11/30	-	-	W,T	F
Evening Grosbeak	6	0	0	2	1	3	10/25-11/3	-	-	T	R
House Sparrow	56	39	68	109	63	279	8/21-11/29	-	-	P	U
Total Species	-	69	102	120	76	145	-	-	-	-	-
Total Individuals	-	7,587	15,416	65,629	20,560	109,192	-	-	-	-	-

[a]145 species, plus *Empidonax* species and one race.
[b]This total is the sum of all days of censusing and in some cases may include duplication of individuals that remained in the study area for several days.
[c]Dates given are for first and last records, or of single occurrences.
[d]Residence status: P—Permanent, T—Transient, S—Summer, and W—Winter.
[e]Relative abundance: A—Abundant, C—Common, F—Fairly Common, U—Uncommon, R—Rare, Ac—Accidental.

TABLE 4. Summary of relative abundance in relation to residence status for 145 species recorded on the 1965 censuses.

Relative Abundance	Number of Species: Residence Status[a] P	P,T	S	S,T	T	W,T	Ac	Total	%
Abundant	1	7	0	1	2	5	-	16	11
Common	1	6	0	3	2	7	-	19	13
Fairly Common	1	3	0	3	18	4	-	29	20
Uncommon	3	1	0	2	24	8	-	38	26
Rare	2	0	2	1	32	1	-	38	26
Accidental	-	-	-	-	-	-	5	5	4
Total Species	8	17	2	10	78	25	5	145	100

[a]Ac = Accidental, P = Permanent, S = Summer, T = Transient, W = Winter.

TABLE 5. Birds banded in fall 1965 at Bodie Island, North Carolina.

Species	September 16	17	19	23	24	26	29	30	October 6	24	26	28	31	November 9	Fall Total
Sharp-shinned Hawk								1				1	1		3
Yellow-billed Cuckoo						1									1
Yellow-bellied Sapsucker						2	2	1	6						11
Downy Woodpecker							2					2	3	1	8
Northern Flicker						4	1	2	5			1			13
Eastern Wood-Pewee							1		1						2
Acadian Flycatcher									1						1
Eastern Phoebe						1									1
Carolina Chickadee		1			2	1									4
Red-breasted Nuthatch							1	3	2		1	1	1		9
White-breasted Nuthatch									1						1
Brown Creeper						2		5	24	3	3		12		49
Carolina Wren								3	1				1		5
House Wren								2					1		3
Winter Wren											1		2		3
Golden-crowned Kinglet											6	2	7		15
Ruby-crowned Kinglet									1		3	2	3		9
Gray-cheeked Thrush						3		2	2						7
Swainson's Thrush						15	1	8	5						29
Hermit Thrush												4	5		9
Wood Thrush									3						3
Gray Catbird	2	8	1	7	25	11	3	16	11	4	3	6	12	3	112
Northern Mockingbird						2					1				3
Brown Thrasher			1		3	1	1	3	1	1					11
White-eyed Vireo	2				1	1									4
Philadelphia Vireo		1													1
Red-eyed Vireo		1				4		1							6
Orange-crowned Warbler														1	1
Northern Parula			1						1		1				3
Magnolia Warbler			1			7	3	5	8						24

Table 5. (continued)

Species	September								October					November	Fall Total
	16	17	19	23	24	26	29	30	6	24	26	28	31	9	
Cape May Warbler			1			17		1		2					21
Black-thr. Blue Warbler					1	9		2	5						17
Yellow-rumped Warbler						5		1	5	8	111	83	210	71	494
Black-thr. Green Warbler						2									2
Blackburnian Warbler						1									1
Pine Warbler						1									1
Prairie Warbler		1				1		1							3
Palm Warbler						4		2	3		1		2		12
Blackpoll Warbler			1			12	2	8	2		1				26
Black-and-white Warbler		1				6		3	1						11
American Redstart	1	3	16	4	5	72	6	21	12	1					141
Ovenbird						1	1	1	2						5
Northern Waterthrush						4									4
Connecticut Warbler						2									2
Common Yellowthroat	1	2	5	4	19	25		14	5	1			1	1	78
Wilson's Warbler						1									1
Yellow-breasted Chat								2							2
Scarlet Tanager						2			3						5
Northern Cardinal					2			1	2				1		6
Rose-breasted Grosbeak						3									3
Indigo Bunting						1									1
Rufous-sided Towhee		2		1	1		1	2		1			3		11
Chipping Sparrow													2		2
Field Sparrow														1	1
Song Sparrow											1	1	2	1	5
Swamp Sparrow											1	8	4	7	20
White-throated Sparrow											2	2	3		7
Dark-eyed Junco						1			1			2	29		33
Red-winged Blackbird					1	4									5
Baltimore Oriole			1	1		1		2							5
Total Individuals	6	20	28	17	60	230	25	113	114	21	136	115	305	86	1,276
Total Species	4	9	9	5	10	36	13	27	27	8	14	13	21	8	60
Max. Number of Nets	10	12	12	12	18	18	11	11	11	6	6	6	24	12	24
Number of Net Hours	45	102	150	66	134	215	47	138	81	20	23	42	161	93	1,317

the fall of 1966, 25 individuals of 23 species were taken in the study area during a period of 10 days (30 September through 9 October) to further support the earlier work.

The banding effort in the fall of 1965 showed the age classes (274 individuals aged by plumage or state of skull ossification) to be 33% adults and 67% immatures. This 1:2 age ratio was also evident separately for both September and October. For birds collected in the fall of 1965 and 1966 (Table 6), the age classes (N = 34 individuals) were 12% adults and 88% immatures. The difference in the age ratios for banding and collecting may be an artifact of the small sample size in the case of migrants collected. The collected material revealed about an equal sex ratio of migrants (N = 35) with 54% males and 46% females. No consistent effort was made to determine age and sex classes on the censuses nor was the sex determined for banded individuals except for the adult males, whose obvious plumage dimorphism made their sex readily apparent.

Altogether 110,482 individuals (accumulative daily totals) of 148 land-bird species were recorded (censusing transects, banding, and miscellaneous observations) in the fall of 1965. Three of these species were not recorded on a census or while banding: a Saw-whet Owl (*Aegolius acadicus*) flew into the side of my automobile and was collected; a single Least Flycatcher (*Empidonax minimus*) was collected; and a Red-bellied Woodpecker (*Melanerpes carolinus*) was heard calling on a day when I was operating the mist nets. These 148 species represented 27 Families of eight Orders (Appendix A). Of the 148 species, 11% were abundant, 13% common, 20% fairly common, 27% uncommon, 25% rare, and 4% accidental.

Based on the total number of individuals recorded (Appendices B and C), 17 species were represented by one individual, 43 species from two to 10 individuals, 25 species from 11 to 50 individuals, 20 species from 51 to 100 individuals, 20 species from 101 to 500 individuals, seven species from 501 to 1,000 individuals, 14 species from 1,001 to 10,000 individuals, and three species by more than 10,000 individuals. The Tree Swallow (*Tachycineta bicolor*) was the most abundant species with a total of 45,467 individuals. The next most abundant species was the Red-winged Blackbird (*Agelaius phoeniceus*) with 13,712 individuals, followed by the Yellow-rumped Warbler (*Dendroica coronata*) with 12,313. A complete ranking of species by the number of individuals recorded is given in Appendix C.

The migration spanned the entire period of observation from 21 August through 31 November. The major portion of the migration took place during the last half of September and through October (Fig. 9). The volume of individuals and number of species increased through September, reached a peak in early October, and slowly decreased from mid-October through November. The peak recorded movement was on 5 October, when 86 species were observed.

Sequence of the Migration

The birds migrated in a definite sequence during the fall. Some migrated early (August to mid-September), some during mid-fall (mid-September to mid-October), others in late fall (mid-October through November), and still others throughout most of the migration period. One or more representative species were chosen for each time period and are presented graphically (Fig. 10-15). Most of the flycatchers and swallows (except the Tree Swallow) and a few of the warblers migrated early; the cuckoos, woodpeckers, thrushes, vireos, and most warblers migrated in mid-fall; and the hawks, owls, and most sparrows and Cardueline finches migrated in late fall.

Among some of the earlier migrants were the Ruby-throated Hummingbird, most of the flycatchers, Bank Swallow, Barn Swallow, White-eyed Vireo, Yellow Warbler, Prairie Warbler, Prothonotary Warbler, Northern Waterthrush, Yellow-breasted Chat, Northern Oriole (Fig. 10), and Lark Sparrow. Some of the migrants of the mid-fall period included the American Kestrel, Yellow-bellied Sapsucker, Tree Swallow, Gray-cheeked Thrush, Swainson's Thrush, Red-eyed Vireo, Magnolia Warbler, Cape May Warbler (Fig. 11), Blackpoll Warbler, and American Redstart (Fig. 12). Included among the late-fall migrants were the Northern Harrier, Sharp-shinned Hawk, Merlin, Peregrine Falcon, Short-eared Owl, House Wren, Winter Wren, Golden-crowned Kinglet, Ruby-crowned Kinglet (Fig. 13), Hermit Thrush, American Robin, Yellow-rumped Warbler, Savannah Sparrow, Swamp Sparrow, White-throated Sparrow (Fig. 14), White-crowned Sparrow, Dark-eyed Junco, and Snow Bunting. Some of the species that were found migrating throughout most of the fall were the Mourning Dove (Fig. 15), Northern Flicker, Red-breasted Nuthatch, Gray Catbird, Brown Thrasher, Cedar Waxwing, Common Yellowthroat, Song Sparrow, Red-winged Blackbird, and Eastern Meadowlark.

Bird Movement Relative to Surface Weather

Twenty cold fronts (Table 7) passed through the study area from August through November 1965: 2 in August, 3 in September, 8 in October, and 7 in November. During this period, precipitation was about 43% below normal, although the trend of distribution was, as usual, highest in August and decreasing through the fall. Wind speeds (Table 8) were about normal except for November, which was 1 knot below normal. No hurricanes or other heavy storms occurred during the period. Sky conditions during the daylight hours of the 102-day period were as follows: clear on 59 days—9 in August, 18 in September, 16 in October, and 16 in November; partly cloudy on 13 days—1 in August, 3

TABLE 6. Specimens collected at Bodie Island, North Carolina, in 1965 and 1966.

Species	Catalogue Number	Age[a]	Sex[b]	Date
Fall 1965				
Northern Saw-whet Owl	NCSM 3194	A	F	25 Oct.
Eastern Wood-Pewee	NCSM 3258	I	F	6 Oct.
Yellow-bellied Flycatcher	NCSM 3248	-	M	30 Sept.
Least Flycatcher	NCSM 3254	I	M	6 Oct.
Swainson's Thrush	NCSM 3370	I	F	5 Oct.
Tennessee Warbler	NCSM 3467	I	F	6 Oct.
Black-throated Blue Warbler	NCSM 3503	-	F	26 Oct.
Common Yellowthroat	NCSM 3604	I	M	24 Sept.
Wilson's Warbler	NCSM 3619	I	M	6 Oct.
White-crowned Sparrow	NCSM 3824	I	F	23 Oct.
White-crowned Sparrow	NCSM 3823	I	M(?)	30 Oct.
Fall 1966				
Yellow-bellied Sapsucker	NCSM 3225	I	M	7 Oct.
Eastern Wood-Pewee	NCSM 3259	I	F	8 Oct.
Western Kingbird	NCSM 3241	I	M	8 Oct.
Red-breasted Nuthatch	NCSM 3306	I	F	2 Oct.
Brown Creeper	NCSM 3313	I	M	8 Oct.
Winter Wren	NCSM 3320	I	M	2 Oct.
Gray-cheeked Thrush	NCSM 3374	I	F	30 Sept.
Swainson's Thrush	NCSM 3371	I	F	1 Oct.
Wood Thrush	NCSM 3360	A	M	8 Oct.
Wood Thrush	NCSM 3361	I	M	8 Oct.
Tennessee Warbler	NCSM 3468	A	M	2 Oct.
Magnolia Warbler	NCSM 3485	I	M	7 Oct.
Magnolia Warbler	NCSM 3486	I	F	8 Oct.
Black-throated Blue Warbler	NCSM 3507	I	M	8 Oct.
Black-throated Green Warbler	NCSM 3521	I	M	8 Oct.
Blackburnian Warbler	NCSM 3530	I	F	2 Oct.
Pine Warbler	NCSM 3558	I	F	8 Oct.
Bay-breasted Warbler	NCSM 3545	I	M	8 Oct.
Blackpoll Warbler	NCSM 3553	I	F	8 Oct.
Northern Waterthrush	NCSM 3582	I	M	8 Oct.
Connecticut Warbler	NCSM 3594	A	F	2 Oct.
Scarlet Tanager	NCSM 3701	I	M	2 Oct.
Clay-colored Sparrow	NCSM 3811	I	F	1 Oct.
Grasshopper Sparrow	NCSM 3781	I	M	9 Oct.
Lincoln's Sparrow	NCSM 3842	I	M	8 Oct.

[a]Age: A—adult; I—immature.
[b]Sex: M—male; F—female.

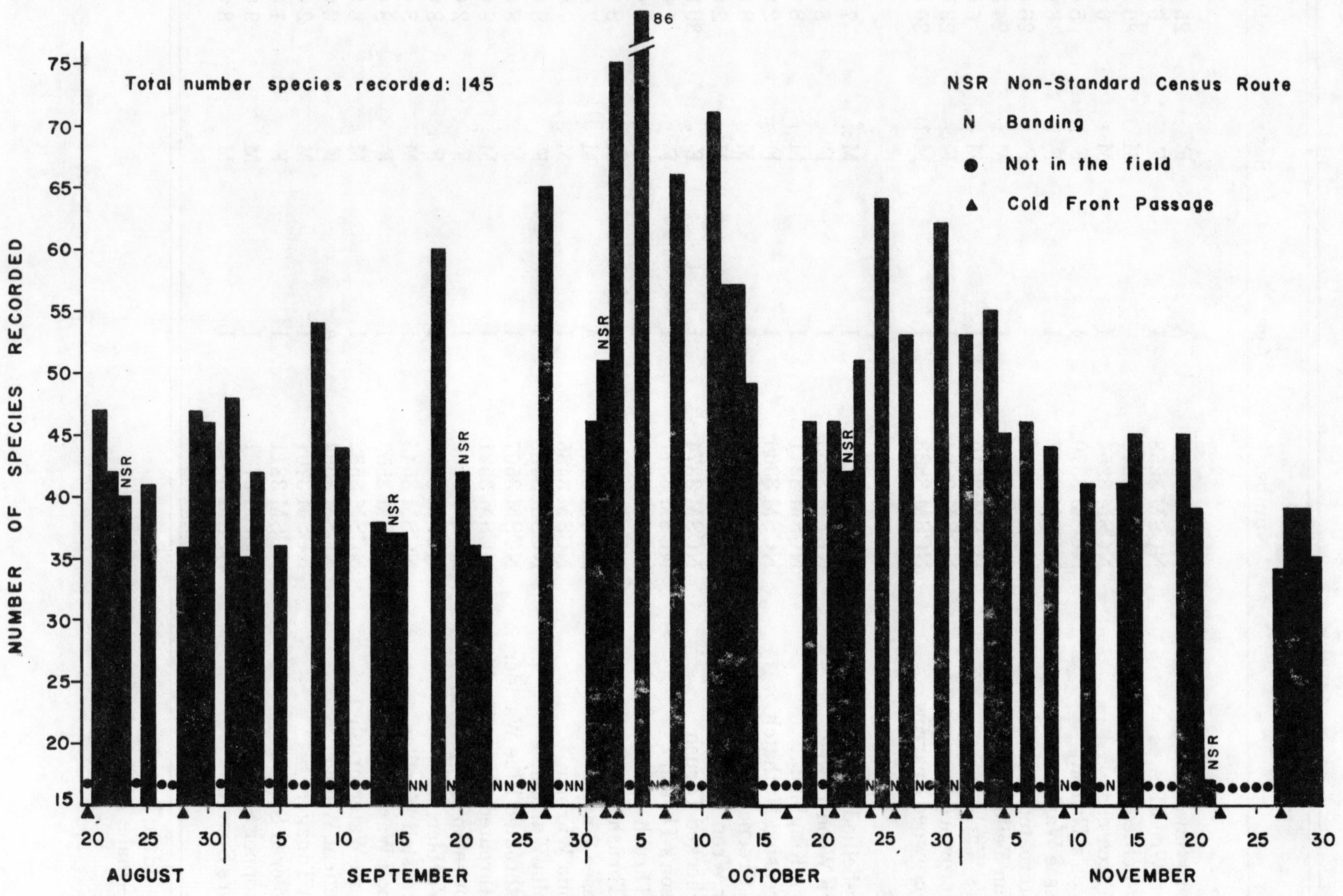

Fig. 9. Total number of land-bird species for each day of censusing, fall 1965.

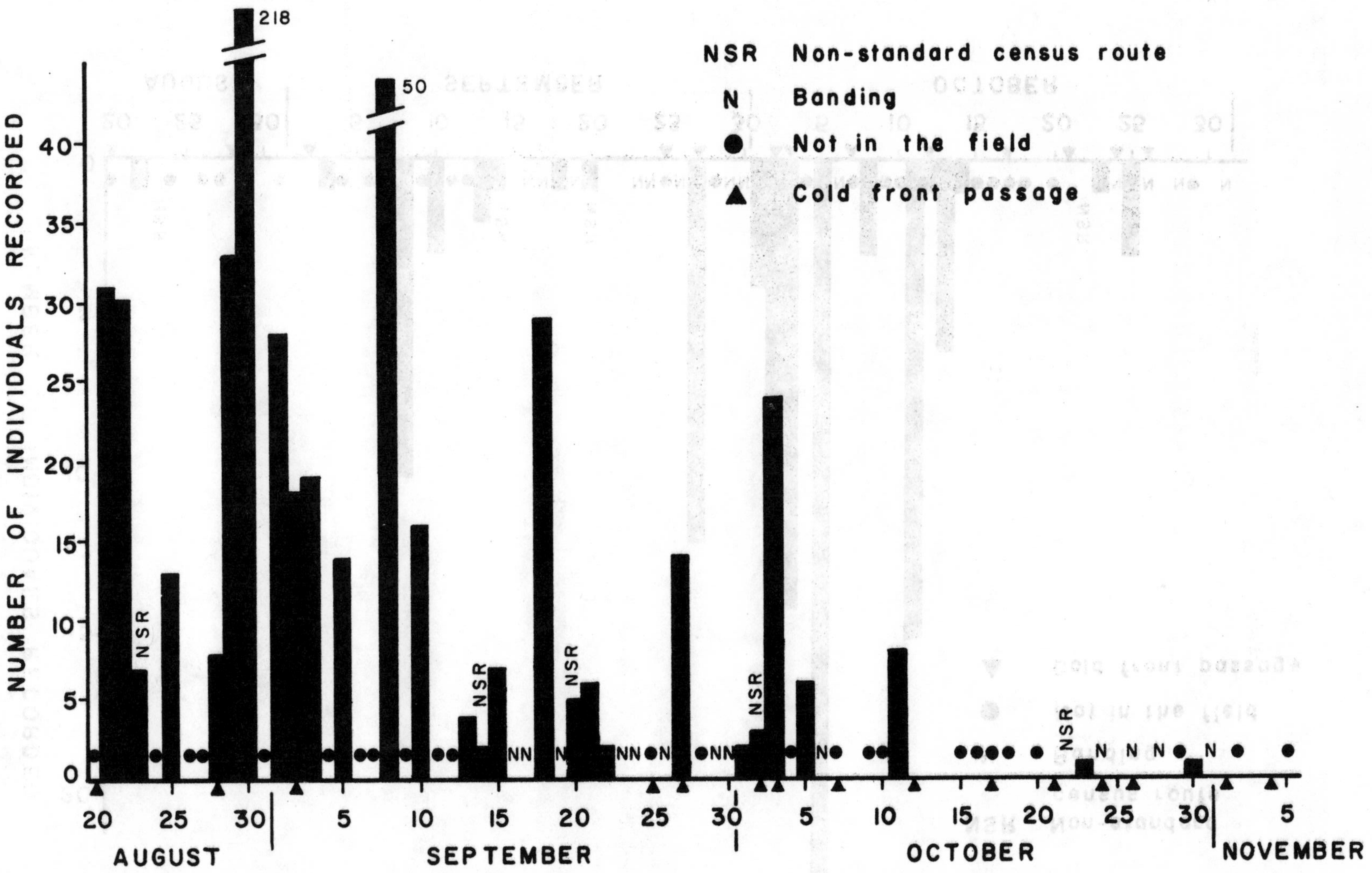

Fig. 10. Total number of Northern Orioles for each day of censusing, fall 1965.

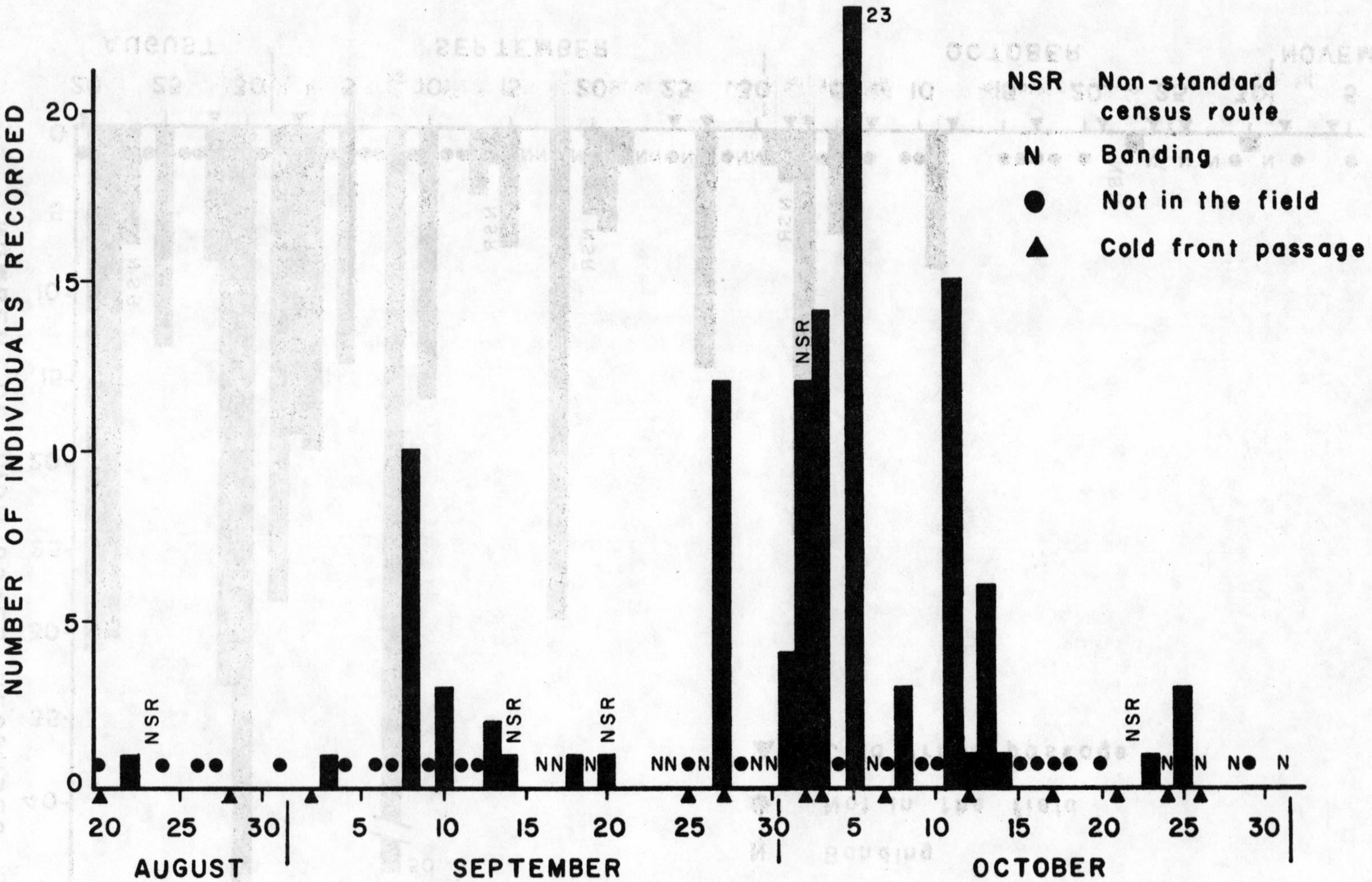

Fig. 11. Total number of Cape May Warblers for each day of censusing, fall 1965.

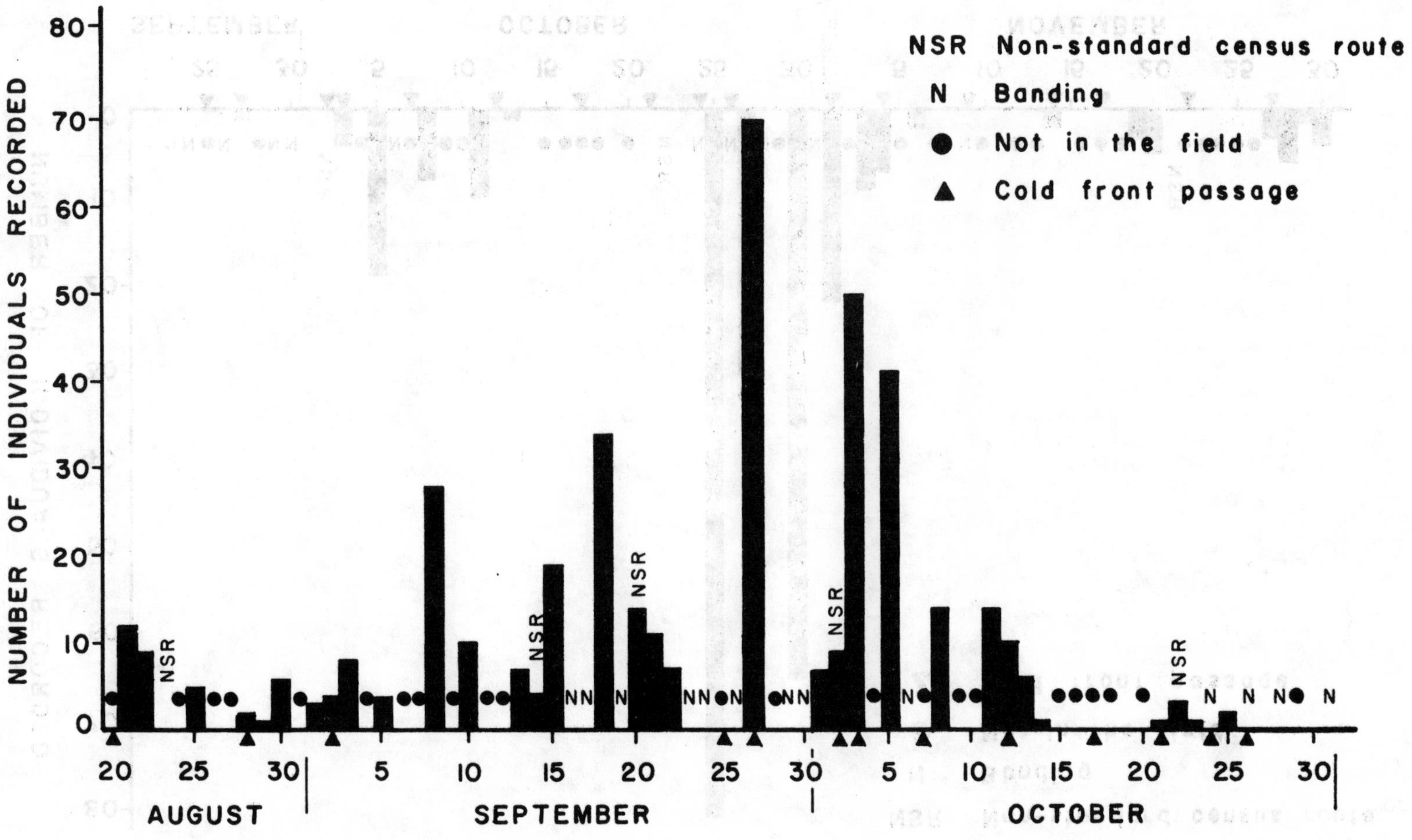

Fig. 12. Total number of American Redstarts for each day of censusing, fall 1965.

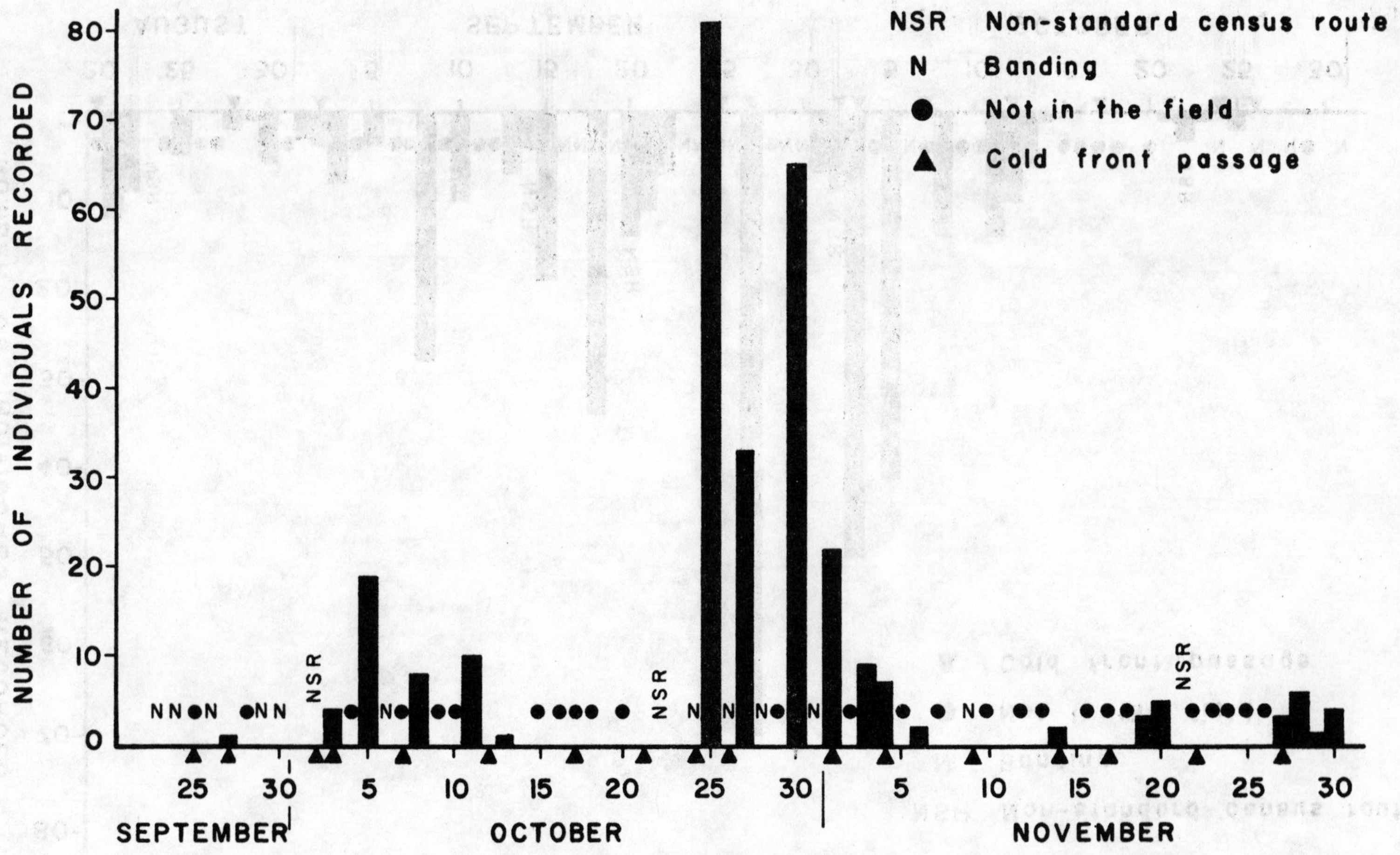

Fig. 13. Total number of Ruby-crowned Kinglets for each day of censusing, fall 1965.

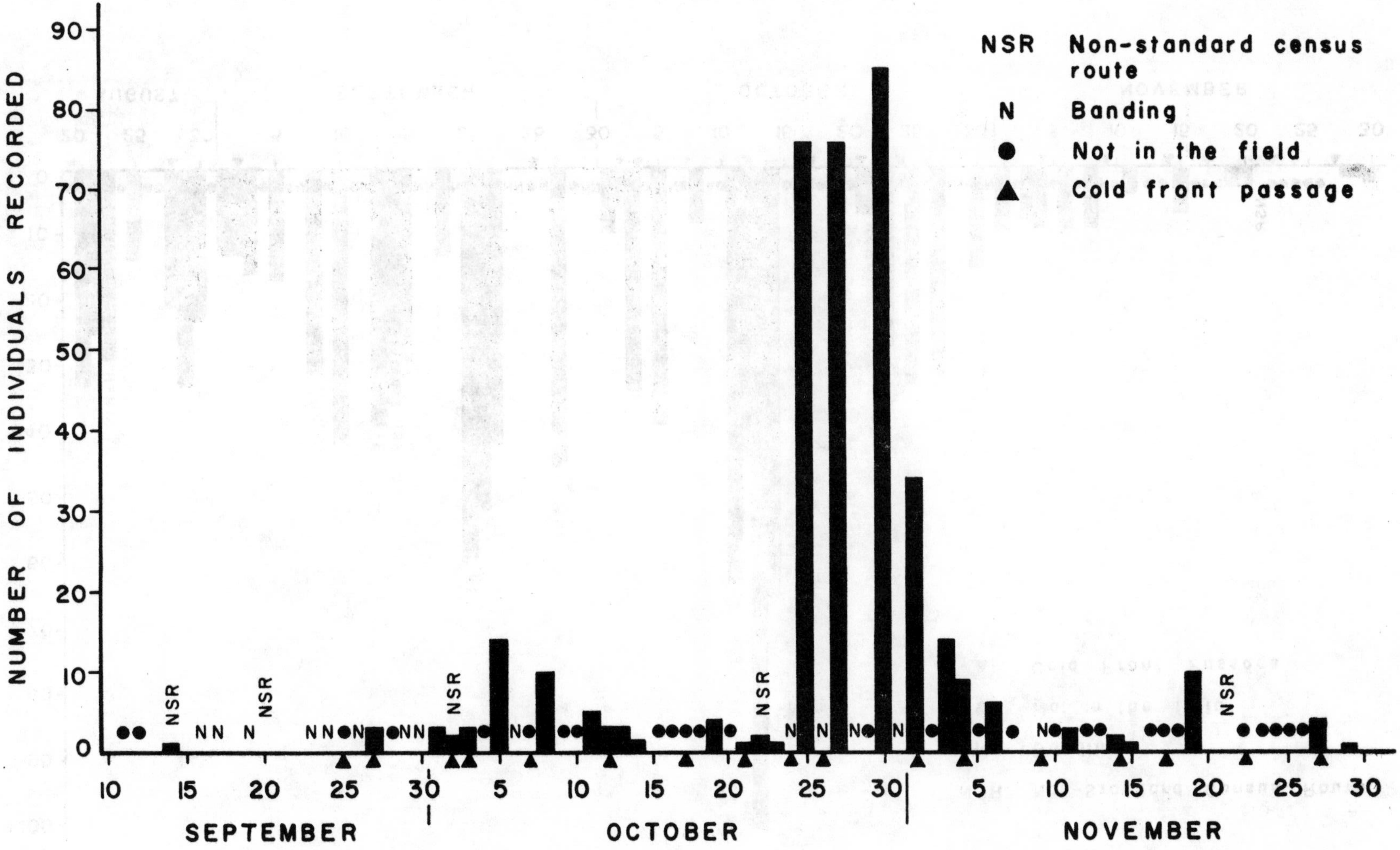

Fig. 14. Total number of White-throated Sparrows for each day of censusing, fall 1965.

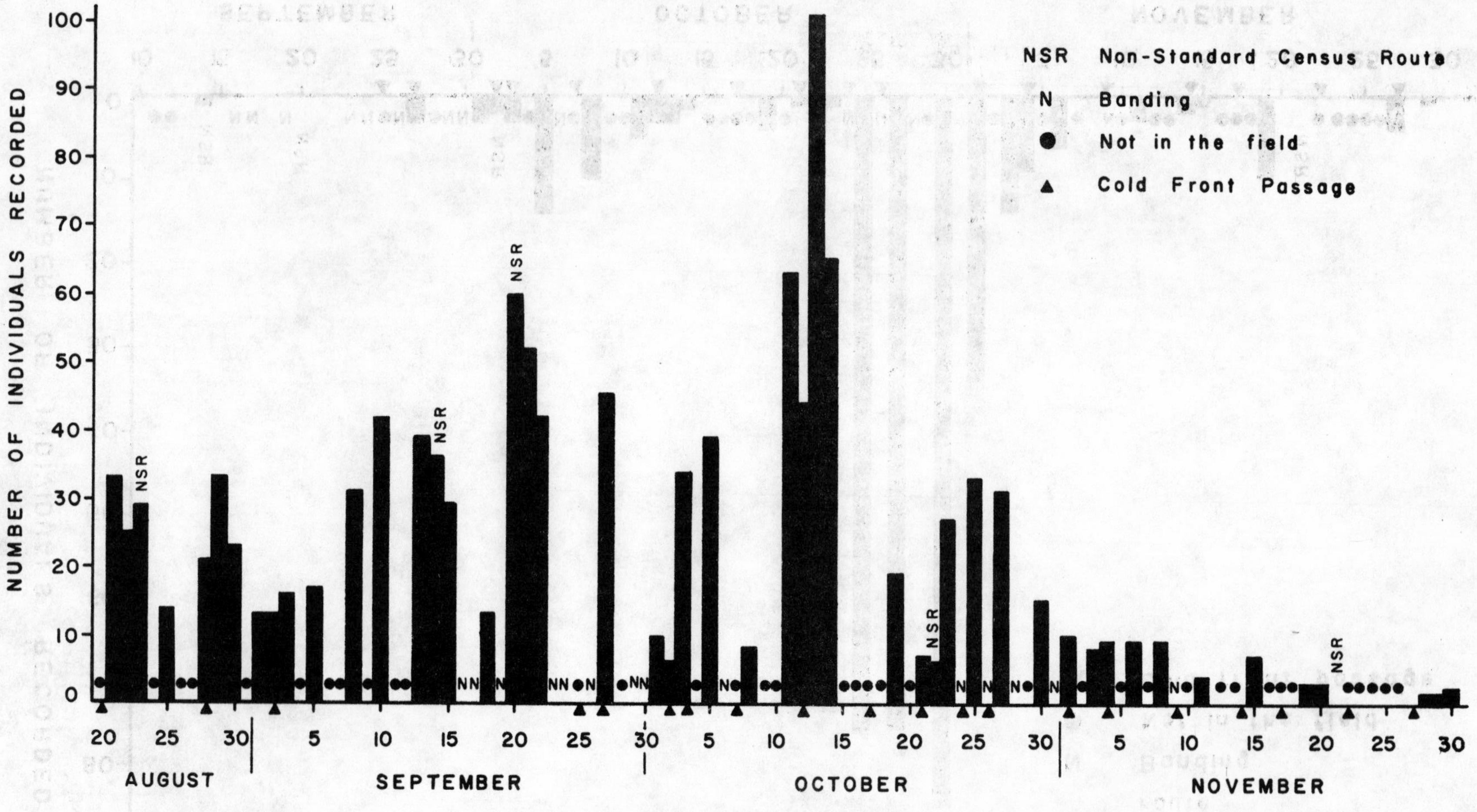

Fig. 15. Total number of Mourning Doves for each day of censusing, fall 1965.

TABLE 7. Cold fronts passing through the study area, fall 1965.[a]

August				September				October				November				
Date	Time		Strength	Date	Time		Strength	Date	Time		Strength	Date	Time		Strength	Total
20	L	a.m.	W	2	E	a.m.	W	2	M	a.m.	S	1	E	a.m.	S	
28	L	p.m.	S	25	M	a.m.	S	3	L	p.m.	S	4	L	p.m.	W	
				27	M	a.m.	W	7	M	p.m.	W	9	L	a.m.	S	
								12	M	p.m.	W	14	M	p.m.	W	
								17	E	p.m.	S	17	M	a.m.	S	
								21	L	a.m.	W	22	L	a.m.	S	
								24	E	a.m.	S	27	M	p.m.	S	
								26	M	a.m.	W					
Total 2				3				8				7				20

[a]Time designations:

	a.m.	p.m.
E—early	0200 ± 2 h	1400 ± 2 h
M—mid	0600 ± 2 h	1800 ± 2 h
L—late	1000 ± 2 h	2200 ± 2 h

The strength of a cold front is based on more than 12°C drop in dew point, plus the clockwise wind shift with speeds of 9 knots or more, but not necessarily occurring at the time of the greatest dew point shift. S—strong front, W—weak front.

in September, 5 in October, and 4 in November; and overcast on 30 days—1 in August, 9 in September, 10 in October, and 10 in November.

The conditions produced by the passage of a cold front prevailed from one to several days. The greatest number of species was recorded on the first or second day following a frontal passage, and diversity decreased rapidly thereafter, with the smallest number of species recorded on the day prior to the passage of an oncoming front. Strong fronts produced a greater number of species than did weak fronts. On the other hand, the largest number of individual birds was recorded, in most cases, on the day of frontal passage, and the strength of the front appeared to have little if any relationship to the number of individuals.

Of the 20 cold fronts recorded during the 102-day period, 11 were classed as strong and nine as weak (Table 7). The strength of a cold front was based on there being more than a 12°C drop in the dew point, plus the clockwise wind shift with speeds of 9 knots or more, but not necessarily occurring at the time of the greatest dew point shift.

The strong fronts in August, September, and October were each followed by a good migratory flight (Fig. 9), whereas in November the fronts appeared to have less effect on the majority of migrants. There was some migration on days following weak cold fronts and also when there was no passage of a frontal system, but on such days migration was usually less pronounced, except for two flights that occurred in September. After 2 September no cold front passed until 24 September, yet a good flight occurred on 8 September and again on 18 September (Fig. 9, 10, 11, 12, 15).

On 8 September, there was a drop of 14°C in the mean temperature from the preceding day, the wind was from a northerly direction with a speed greater than 10 knots for several days preceding, 0.38 mm (0.15

inches) of rain was recorded on 7 September, and the sky conditions during the preceding 24 hours were clear. The conditions were somewhat different on 18 September. There was little change in the temperature for the preceding 48 hours. The winds shifted from northwest at 5 knots at 8 a.m. to southeast at 7 knots at 4 p.m. on 17 September and to north-northwest at 6 knots at 8 a.m. on 18 September. There was no precipitation during the 3-day period prior to 18 September, and the sky conditions were mostly clear for several days preceding the eighteenth.

The fronts that occurred during October produced the largest number of species and the greatest number of migrants (Fig. 9). This was also the case for other years.

General Habitat Usage

There was a pronounced difference in use of the available habitats by the migrants. Based on general observations, ranking of the major habitats, from the most heavily used to the least used, was as follows: (1) pine thicket, (2) herb-shrub and shrub thicket, (3) dunes, (4) fresh marsh, (5) high marsh, (6) low marsh, and (7) beach. The pine thickets were quite distinct from the lower, surrounding vegetation; this habitat tended to attract most species in good numbers. The shrub thickets and herb-shrub habitats were used about equally by many small land birds, but not to the extent of the pine thickets. The remaining five habitats were used by only a few species. Only an occasional Palm Warbler or Savannah Sparrow was noted on the upper portion of the beach. Diurnal migrants, such as hawks, nighthawks, swifts, kingbirds, and swallows, flew over all of the habitats, moving southward down the barrier beaches.

DISCUSSION

Censusing was the major method of sampling; mist-netting and collecting were secondary. The census technique was used mainly because I worked alone most of the time. With this method the broad spectrum of migration could be studied, whereas with banding one person could not operate a sufficient number of mist nets to give comparable results. Furthermore, many species are not readily taken in mist nets. These include the falcons, Buteos, large owls, swifts, kingbirds, swallows, crows, and some of the blackbird group. In censusing, a large area and most of the major habitats could be worked in one day. On the other hand, mist-netting gave better quantitative data on the secretive species. However, no species was netted that was not also recorded on the censuses.

Several features of the study area greatly facilitated the field work. The small land mass tended to concentrate the migrants. The relatively limited amount of upland habitats further concentrated many of the migrants and made coverage of these habitats much easier. The low height of the vegetation, 0.6 to 4.6 m (2-15 feet) in most cases, made observations easier than under conditions found on the mainland where many of the migrants would be in the tops of trees 15 to 37 m (50-120 feet) tall. The low vegetation also permitted observation of birds flying overhead for a considerable distance. The intermingling of the different habitats provided extensive ecotonal conditions attractive to most species of small birds.

The barrier islands of northeastern North Carolina are well within the usual fall migration corridor of many of the eastern North American land birds. The occurrence of large numbers of migrants, particularly nocturnal migrants, along the coast of the northeastern United States in autumn has been thought to be the result of birds drifted by winds from more inland routes (Allen and Peterson 1936; Stone 1937; Baird and Nisbet 1959, 1960). This appears to be the case for the study area, because 57% of the 148 species recorded in the fall of 1965 were either uncommon, rare, or accidental and much of the heaviest movement occurred when winds were from a westerly direction. On the nearby mainland 80 to 90% of the species in these three categories would be expected during migration in substantial numbers. Also, a number of species that occur on the adjacent mainland as regular migrants were not recorded in the study area.

In general, the magnitude of the fall migration through the Bodie Island-Pea Island area was found to be comparable with results from Operation Recovery stations at localities along the Atlantic coast from New England south to Virginia (Baird et al. 1958; Baird and Nisbet 1959; Scott 1963, 1965; Smith 1966). However, a few species did not occur on the North Carolina barrier islands,

TABLE 8. Wind directions and speeds (in knots), fall 1965.

	August		September		October		November	
Date	8 a.m.	4 p.m.	8 a.m.	4 p.m.	8 a.m.	4 p.m.	8 a.m.	4 p.m.
1	SE 8	SSW 16	SSW 12	SSW 17	S 12	SW 15	NNW 13	NNE 11
2	SSW 10	SW 16	NE 18	NE 17	NNE 24	NE 14	NE 14	NE 14
3	NNE 6	ESE 10	NE 18	NE 14	WNW 10	W 12	SSW 6	SSW 12
4	SSE 13	SSW 11	ENE 19	ENE 20	NNE 20	N 10	WSW 11	WSW 13
5	NNE 6	E 10	NE 18	NE 14	NNW 11	NNW 9	ENE 25	NE 22
6	SE 7	S 12	NE 18	NE 17	NW 6	SE 10	ENE 12	ENE 10
7	WNW 4	SSE 11	ENE 14	NE 11	SSE 17	S 17[b]	ENE 6	ESE 6
8	SW 5	SSW 11	NNE 8	ESE 8	WSW 16	SW 17	ENE 4	SSW 12
9	WSW 8	SW 19	NNW 7	SSW 12	W 10	SSW 13	WNW 8	NE 23
10	SW 10	SW 21	WSW 8	SW 17	NW 11	NW 15	NE 18	SSW 12
11	NW 7	E 8	WSW 10	SSW 9	NW 7	ESE 5	ENE 18	NE 18
12	NE 8	SE 10	SSW 12	SSW 13	W 9	SW 7	NE 20	NE 20
13	WNW 10	SW 16	SSW 10	SSW 15	NE 10	ENE 12	NE 20	NNE 11
14	NW 10	ESE 5	W 10	ENE 5	ESE 10	SE 9	NW 5	NE 13
15	NE 9	NE 7	W 8	SW 14	ESE 10	ESE 15	NE 10	NE 6
16	- -	ENE 7	W 10	SSW 14	NNE 12	NE 18	WSW 6	WSW 6
17	- -	E 7	NW 5	SE 7	NE 23	NE 28	W 20	NW 20
18	S 4	SSW 10	NNW 6	SW 5	ENE 22	ENE 24	NW 21	N 15
19	WSW 14	SW 16	N 7	ESE 6	ENE 17	ENE 18	N 6	N 5
20	N 10	NE 13	E 4	SE 7	ENE 21	ENE 16	ENE 10	ENE 8
21	ENE 14	E 11	ESE 2	SSE 10	NE 17	NW 9	SW 6	SSW 5
22	SW 7	SSW 15	SSE 9	SSE 12	ESE 6	SW 10	SW 9	NNW 15
23	W 13	E 35[a]	SE 6	SE 12	WSW 6	W 12	NW 10	WSW 15
24	N 4	NNE 12	WSW 7	SE 20	NW 12	N 18	NNW 9	N 9
25	E 9	SE 11	N 14	NNE 18	NNW 15	NNE 4	NE 4	E 8
26	W 11	SW 13	NE 9	NE 9	NW 8	SW 18	ESE 5	ESE 5
27	WNW 10	ESE 9	NE 15	NE 15	NNW 8	ENE 8	SW 10	SW 15
28	SW 16	SW 22	ENE 25	ENE 21	NW 8	NE 12	NW 7	SW 15
29	NE 22	NNE 9	NE 18	NNE 13	NNE 25	NNE 14	N 13	NW 8
30	NNE 8	NE 12	ENE 11	E 9	NW 6	SSE 9	NE 11	NNE 25
31	ESE 9	ESE 8			W 9	SW 14		
Mean Speed 1965	11		12		13		12	
Normal[c]	11		12		13		13	

[a]Gusting to 58 knots.

[b]Gusting to 55 knots.

[c]Data from Carney and Hardy (1964).

such as the Broad-winged Hawk (*Buteo platypterus*), while others, such as Veery (*Catharus fuscescens*), Gray-cheeked Thrush (*C. minimus*), and Swainson's Thrush (*C. ustulatus*), occurred in very small numbers compared to other coastal localities to the north (i.e. Cape May, New Jersey; Ocean City, Maryland; and Kiptopeke, Virginia). Had a continuous banding program been in operation at Bodie Island, the number of individuals recorded of the Subfamily Turdinae and other secretive species probably would have been much higher (see Stone 1937; Scott 1963, 1965; Smith 1966).

Although the study area appears to be slightly east of most fall land-bird movements in northeastern North Carolina, an impressive number of birds pass through the area each autumn. In 1965, I recorded 110,482 individuals of 148 species of land birds; and data collected in the fall of 1963, 1964, and 1966 further substantiate significant migratory movements. Thus, the species richness and magnitude of migratory movement through the barrier islands in 1965 are believed to be the norm.

The period of fall migration of land birds on the barrier islands of North Carolina begins in mid-July and extends to mid-December. The extreme dates were determined from additional work in 1966 and periodic observations through 1984, as published in *The Chat* and *American Birds*. July migrants were the swallows, Black-and-white Warbler, American Redstart, Prothonotary Warbler, Yellow-breasted Chat, and Common Grackle. This early migratory movement is not readily observed on the mainland of North Carolina and in general was not recorded in the literature until the 1970s. The major portion of the migration was from mid- to late-August through November. The later migratory movement primarily involved the Emberizinae.

The sequence of migration was readily observed, with different species migrating at different periods in the fall. There seemed to be four major groupings of migrants: (1) those that migrated early, (2) those that migrated in mid-fall, (3) those that migrated late, and (4) those that migrated over most of the fall period. The mid-fall category contained the largest number of species, and the extended fall period the fewest species. In general, taxa of Neotropical affinity tended to migrate earlier in the fall than those of the Nearctic.

A number of factors, singly or in combination, that may influence the sequence of fall migration among land birds are: difference in the time of departure from the breeding grounds, distance that must be traveled from the breeding to the wintering grounds, the amount of preferred food available at the time of the migration, and weather. The time at which the breeding cycle is completed or young are independent of the parents, and the necessary physiological state reached for the departure from the breeding grounds, differs widely among species. Those species that travel the greatest distances from the breeding to the wintering grounds (e.g. the Purple Martin) are usually earlier migrants than those that travel a shorter distance (e.g. most Emberizinae). There may be a relationship between the time a given species migrates and the availability of its preferred food, particularly among the insectivores. If all or most species migrated at the same time through a given area, certain items of food might be inadequate to sustain the large influx of birds. Thus, it appears that several factors may govern the sequence of migration among the different species.

Assuming the birds to be in the necessary physiological state, surface weather appears to have a significant effect on migratory movement. As had been found by most workers, heavy migration in the fall of 1965 corresponded best with the passage of a cold front.

Except for the last two fronts in late November 1965, all the strong fronts had surface winds from the western or northern quadrants. Winds from the west and northwest would aid migrants in reaching the North Carolina barrier islands in Dare County or perhaps be the reason for most of them being on the barrier islands rather than on the mainland. The largest number of species recorded during the fall of 1965 is a case in point. On 5 October, 86 species were recorded following the frontal passage of 3 October with winds from the west-northwest and west. This front had been preceded by a strong front on 2 October with winds from the north-northeast and northeast.

Relatively little work has been done on habitat selection during fall migration. Stewart and Robbins (1958) presented a general description of habitats for many land-bird migrants in Maryland. Parnell (1964) made a detailed study of habitat relations of migrating spring warblers in the lower piedmont of North Carolina. Most other habitat studies have been done during the breeding and wintering seasons. On the barrier islands of North Carolina, I found that many of the fall migrants were out of what would appear to be their usual habitat. Often the usual habitat of a species was not available in the region, forcing it to use one or more alternate habitats. A few examples of species observed out of the usual habitat are: Yellow-bellied Sapsucker, Cedar Waxwing, and Solitary Vireo in pine thickets; Downy Woodpecker, Hairy Woodpecker, Yellow-bellied Flycatcher, Great Crested Flycatcher, Red-breasted Nuthatch, Brown Creeper, kinglets, thrushes, Red-eyed Vireo, Black-and-white Warbler, American Redstart, Prothonotary Warbler, Canada Warbler, Rose-breasted Grosbeak, Fox Sparrow, Brown-headed Cowbird, and Northern Oriole in the herb-shrub and shrub thickets; and Palm Warbler and Northern Waterthrush in fresh, high, and low marshes.

A number of permanent resident species of northeastern North Carolina were not recorded on Pea or Bodie Islands in the fall of 1965. Included in this group were Black Vulture (*Coragyps atratus*), Northern Bobwhite (*Colinus virginianus*), Eastern Screech-Owl (*Otus asio*), Great Horned Owl (*Bubo virginianus*), Barred Owl (*Strix varia*), Chuck-will's-widow (*Caprimulgus carolinensis*), Pileated Woodpecker (*Dryocopus pileatus*), Blue Jay (*Cyanocitta cristata*) [now known to occur as an occasional migrant (Harry E. LeGrand Jr., pers. comm.)], Tufted Titmouse (*Parus bicolor*), Brown-headed Nuthatch (*Sitta pusilla*), and Eastern Bluebird (*Sialia sialis*). The lack of suitable habitat and small land mass perhaps best explain the absence of most of these species in the study area.

Among the summer residents of eastern North Carolina that were not recorded are Yellow-throated Vireo (*Vireo flavifrons*), Swainson's Warbler (*Limnothlypis swainsonii*), Louisiana Waterthrush (*Seiurus motacilla*), and Orchard Oriole (*Icterus spurius*). The Lousisiana Waterthrush and Orchard Oriole probably migrate through the study region in July and August. Thus, they may have already moved south before the field work began on 21 August. The other two species probably migrate through the study region in very small numbers.

Transients that probably occurred in the study region in the fall but were not recorded are Broad-winged Hawk, Traill's Flycatcher complex of Alder (*Empidonax alnorum*) and Willow (*E. traillii*) Flycatchers, Golden-winged Warbler (*Vemivora chrysoptera*), Worm-eating Warbler (*Helmitheros vermivorus*), Kentucky Warbler (*Oporornis formosus*), Mourning Warbler (*Oporornis philadelphia*), and Henslow's Sparrow (*Ammodramus henslowii*). These species, except for Alder and Willow Flycatchers, were probably represented during migration by very small numbers of individuals. The Alder and Willow Flycatchers are believed to occur regularly; but because they are almost impossible to identify in the field in the fall except by their call notes, they would have been included with the other members of the genus *Empidonax* that were not identified to the species level. The Cerulean Warbler (*Dendroica cerulea*) occurs infrequently; one bird was observed on Bodie Island on 3 September 1964.

Two species occasionally found during the winter in small numbers on the barrier islands, but not recorded during the fall of 1965, are the Horned Lark (*Eremophila alpestris*) and the Lapland Longspur (*Calcarius lapponicus*). These two species possibly occurred in the study area after the field work had terminated on 30 November.

Some species tend to avoid the barrier islands in their fall migration, using a more inland route. Included among this group are Black Vulture, Turkey Vulture, Cooper's Hawk, Red-shouldered Hawk, Broad-winged Hawk, Red-tailed Hawk, Common Nighthawk, Chimney Swift, Blue Jay, Blue-gray Gnatcatcher, Eastern Bluebird, Veery, Loggerhead Shrike, Yellow-throated Vireo, Warbling Vireo, Yellow-throated Warbler, Worm-eating Warbler, Kentucky Warbler, Hooded Warbler, Canada Warbler, Rusty Blackbird, and Common Grackle.

The above species were not recorded in the Bodie Island-Pea Island area or were recorded in very small numbers. On the mainland to the west and at locations farther inland, however, they are of regular occurrence, and most are common to abundant.

In any population of migratory birds there are a few individuals—just a fraction of a percent, immatures in particular—that for some reason (slight physiological imbalance, misorientation, navigational error, inexperience, mirror-image misorientation [see DeSante 1973], or a combination of these factors) do not follow the species' usual migratory corridor. Acting as independent entities, a few disperse in compass directions different from those of their species, and this tendency, coupled with favorable meterological conditions, is responsible for geographic displacement of such individuals, often hundreds or thousands of kilometers from their usual range. Thus, a few western species are found along the Atlantic coast, and a few eastern species turn up on the Pacific coast. There is now a large volume of such displacement records in *American Birds*, *Auk*, *Condor*, *Wilson Bulletin*, and many other journals.

In recent years there have been numerous reports in the ornithological literature of western species in North Carolina, particularly in immediate proximity to the ocean. Five such western accidentals were found in the study area in the fall of 1965: an immature Swainson's Hawk (*Buteo swainsoni*) at the north end of Pea Island (Potter and Sykes 1980), a Sage Thrasher (*Oreoscoptes montanus*) at Whalebone, a Sprague's Pipit (*Anthus spragueii*) and a Western Meadowlark (*Sturnella neglecta*) in the "goose pasture" on the west side of N.C. 12 on Pea Island in what is now known as New Field, and a Brewer's Blackbird (*Euphagus cyanocephalus*) on Bodie Island at the lighthouse.

The Swainson's Hawk, Sage Thrasher, and Western Meadowlark were the first reports of these species in North Carolina. The Sprague's Pipit was the second record for North Carolina, the first being a bird seen by a number of observers at Chapel Hill (Smith 1959). As of 1966 the Brewer's Blackbird had been recorded only a few times previously in the state, but it now occurs almost annually in migration or in winter. For details on accidentals, refer to Appendix D.

The question arises, have these vagrants from western North America been occurring along the Atlantic Coast over the years? In my opinion the answer is yes. The ocean acts as a barrier to most land-bird migrants, in particular those that are displaced, and hence areas in immediate proximity to the ocean have a higher concentration of displaced migrants than do sites several miles inland. Because of this apparent coastal phenomenon of displaced migrants, many people are searching such areas for rarities. Also, there has been a quantum leap in developing better skills at identifying birds in the field in the last 20 years or so, and selected coastal areas are receiving more thorough coverage than in the past; therefore, more vagrants are being found.

The published literature on the occurrence of land birds in the Bodie Island-Pea Island area of North Carolina has increased considerably during the 20 years that have elapsed since my study in the mid-1960s. New species, such as the Vermilion Flycatcher (*Pyrocephalus rubinus*) (E. LeGrand 1981), have been added to the list of coastal vagrants, and systematic field studies have been conducted elsewhere on the North Carolina coast (e.g. Davis and Parnell 1983), further documenting the influence of cold fronts on the movements of land birds, particularly the nocturnal migrants, southward in the autumn. Recording the presence of land birds is now a well-established routine for the many bird students who briefly visit coastal North Carolina primarily to search for waterfowl, shorebirds, pelagic species, and the anticipated "rarities." I hope that this publication will inspire further systematic field studies of the fall land-bird migration in coastal North Carolina—the Outer Banks in particular—and perhaps elsewhere along the coast of the southeastern United States. The data herein presented are but the tip of the "iceberg" in this diverse, dynamic coastal region molded by the sea.

ACKNOWLEDGMENTS

This study was conducted in partial fulfillment of the requirements for the Master of Science degree in the Department of Zoology at North Carolina State University, Raleigh, N.C., in 1967 under the direction of a Graduate School Advisory Committee composed of Arthur W. Cooper, F. Eugene Hester, and Thomas L. Quay (chairman). At the Cape Hatteras National Seashore, Bruce W. Black, Karl T. Gilbert, Douglas Morris, R. K. Rundell, and Anthony E. Stark arranged for access and lodging at Bodie Island and were responsible for a most pleasant stay in the area. Members of the National Park Service staff aided with clerical matters, Dennis E. McGinnis made available the facilities at the maintenance shop, and J. R. Butler made the mist-net holders. At the Pea Island National Wildlife Refuge, William C. Good and Buster Phillips made the necessary arrangements for access to the closed sections of the refuge and gave encouragement to the study.

The following persons assisted with the field work: Stephen D. Fretwell, Armeda F. Fretwell, Clay L. Gifford, John H. Grey Jr., Robert J. Hader, David R. Hayes, James H. Hunt, Frederick L. Johns, H. Lee Jones, Julian C. Meadows III, James F. Parnell, M. B. Peacock, Elizabeth D. Peacock, Jack Potter, Eloise F. Potter, Thomas L. Quay, Robert K. Smith, Joan J. Sykes, and Philip H. Warren. Arthur W. Cooper and William D. Hood made aerial photographs available for the Bodie and Pea Island areas, and Albert V. Hardy assisted with some of the weather data. Roxie C. Laybourne of the U.S. National Museum verified the identification of some of the specimens collected in 1965 and 1966, and Roger L. Banks assisted with the figures. The study was supported in part by a grant in 1965 from the Josselyn Van Tyne Memorial Fund of the American Ornithologists' Union.

Sincere appreciation is expressed to the above persons and organizations for their encouragement and assistance. I wish to express my special thanks to my long-time friend and mentor, Thomas L. Quay, for his advice, guidance, and effervescent dynamic enthusiasm. It was he who was instrumental in my deciding to undertake this study. I am grateful to John B. Funderburg Jr., Sidney A. Gauthreaux Jr., and Harry E. LeGrand Jr. for review of the manuscript and their helpful comments for its improvement. Finally, I thank my wife Joan for her untiring support, patience, encouragement, and sacrifice during my long absences in the course of the study.

LITERATURE CITED

Allen, R. P., and R. T. Peterson. 1936. The hawk migrations at Cape May Point, New Jersey. Auk 53:393-404.

American Ornithologists' Union. 1983. Check-list of North American Birds. Sixth edition. Amer. Ornithol. Union, Lawrence, Kan.

Baird, J., C. S. Robbins, A. Bagg, and J. V. Dennis. 1958. 'Operation Recovery'--the Atlantic coastal netting project. Bird-Banding 29:137-168.

Baird, J., and I. C. T. Nisbet. 1959. Observations of diurnal migration in the Narragansett Bay area of Rhode Island, in fall of 1958. Bird-Banding 30:171-181.

Baird, J., and I. C. T. Nisbet. 1960. Northward fall migration on the Atlantic coast and its relation to offshore drift. Auk 77:119-149.

Bennett, H. G. 1952. Fall migration of birds at Chicago. Wilson Bull. 64:197-220.

Bishop, L. B. 1901. The winter birds of Pea Island, North Carolina. Auk 18:260-268.

Brimley, C. S. 1942. The status of the wood warblers in North Carolina. Chat 6:40-43.

Burleigh, T. D. 1937. Bird life on the North Carolina coast. Auk 54:452-460.

Carney, C. B., and A. V. Hardy. 1964. Weather and Climate in North Carolina. Bull. 396. Agr. Exp. Sta., North Carolina State Univ. at Raleigh.

Craighill, F. H. 1939. Nag's Head area. Chat 3:79.

Craighill, F. H. 1940. Pea Island in early September. Chat 4:111.

Craighill, F. H., and J. H. Grey Jr. 1938. Nag's Head region. Chat 2:12.

Crosson, D. F., and R. A. Stevenson Jr. 1956. Hawk migrations along the middle eastern seaboard, Delaware to North Carolina. Chat 20:2.

Davis, H. T. 1958. Banding on the upper North Carolina coast. Chat 22:27.

Davis, R. J., and J. F. Parnell. 1983. Fall migration of land birds at Fort Fisher, New Hanover County, N.C. Chat 47:85-95.

Deignan, H. G. 1951. Bird notes from the banks of Dare County, North Carolina. Chat 15:14.

Dennis, J. V., and L. J. Whittles. 1955. The riddle of fall migration at Nantucket. Bull. Mass. Audubon Soc. 39:318-324, 385-394.

DeSante, D. F. 1973. An analysis of the fall occurrences and nocturnal orientations of vagrant wood warblers (Parulidae) in California. Unpubl. Ph.D. dissertation. Stanford Univ., Palo Alto, Calif.

Drury, W. H., Jr., I. C. T. Nisbet, and R. E. Richardson. 1961. The migration of angels. Natur. Hist. (J. American Mus. Natur. Hist.) 70(8):11-17.

Dunbar, G. S. 1958. Historical Geography of North Carolina Outer Banks. Coastal Studies Series No. 3. Louisiana State Univ. Press, Baton Rouge.

Duvall, A. J. 1937. Birds observed on the coast of Virginia and North Carolina. Auk 54:461-463.

Gauthreaux, S. A., Jr. 1978. Migratory behavior and flight patterns. Pages 23-50 in Impacts of Transmission Lines on Birds in Flight. ORAV-142, Proc. Conf. January 31 - February 2, 1978, Oak Ridge Assoc. Univ., Oak Ridge, Tenn.

Green, E., Jr. 1939. The birds of Cape Hatteras. Chat 3:1-20.

Grey, J. H., Jr. 1941. The breeding birds of Pea Island. Chat 5:50-55.

Grey, J. H., Jr., R. Miller, J. Thompson, and R. H. Siler. 1964. Mist-netting on the Outer Banks September 1964. Chat 28:141-142.

Hailman, J. P., and J. J. Hatch. 1964. Mist-netting on the Outer Banks in the fall of 1962, with a banding recovery from Florida. Chat 28:28-29.

Johnson, J. M., J. T. Nichols, and L. Griscom. 1917. Notes from North Carolina. Auk 34:219-220.

Lack, D. 1960. The influence of weather on passerine migration: a review. Auk 77:171-209.

Lack, D. 1963. Migration across the southern North Sea studied by radar. Part IV. Autumn. Ibis 105:1-54.

Lanyon, W. E. 1957. The Comparative Biology of the Meadowlarks (Sturnella) in Wisconsin. Publ. Nuttall Ornith. Club. No. 11. Cambridge.

LeGrand, E. 1981. First record of Vermilion Flycatcher in North Carolina. Chat 45:45.

LeGrand, H. E., Jr. 1981. The relationship of the nocturnal bird migration to the diurnal bird populations in spring and fall in northwestern South Carolina. Ph.D. dissertation, Clemson Univ., Clemson, S.C.

Lincoln, F. C. 1950. Migration of Birds. U.S. Dept. Interior, Fish and Wildlife Serv. Circ. No. 16. Washington, D.C.

Parnell, J. F. 1964. Analysis of habitat relations of the Parulidae passing through the Raleigh, North Carolina, region during the spring migration. Ph.D. dissertation, Dept. of Zool., North Carolina State Univ., Raleigh. Univ. Microfilm, Ann Arbor.

Peacock, E. D. 1964. Some distribution records from the Outer Banks of North Carolina through mist-netting. Chat 28:139-141.

Potter, E. F., and P. W. Sykes Jr. 1980. A probable winter record of Swainson's Hawk from Tyrrell County, N.C., with comments on a fall 1965 sighting from the Outer Banks. Chat 44:76-78.

Pettingill, O. S., Jr. 1956. A Laboratory and Field Manual of Ornithology. Third edition. Burgess Publishing Co., Minneapolis.

Quay, T., and V. Quay. 1939. Pea Island field trip. Chat 3:60-62.

Quay, T. L. 1959. The birds, mammals, reptiles, and amphibians of Cape Hatteras National Seashore Recreational Area. Project Completion Report, National Park Serv., Manteo, N.C.

Radford, A. E., H. E. Ahles, and C. R. Bell. 1964, Guide to the Vascular Flora of the Carolinas. University of North Carolina Book Exchange, Chapel Hill.

Richardson, W. J. 1972. Autumn migration and weather in eastern Canada: a radar study. Amer. Birds 26:10-16.

Richardson, W. J. 1978. Timing and amount of bird migration in relation to weather: a review. Oikos 30:224-272.

Scott, F. R. 1963. Operation recovery at Kiptopeke Beach, Virginia. Raven 34: 53-56.

Scott, F. R. 1965. The 1964 operation recovery at Kiptopeke Beach. Raven 36:71-74.

Simpson, T. W. 1954. The status of migratory hawks in the Carolinas. Chat 18:15-21.

Smith, G. A. 1959. Sprague's Pipit at Chapel Hill. Chat 23:89.

Smith, W. P. 1966. Kiptopeke diary, 1965. Raven 37:7-20.

Stewart, R. E., and C. S. Robbins. 1958. Birds of Maryland and the District of Columbia. North American Fauna No. 62. U.S. Fish and Wildlife Serv., Washington, D.C.

Stone, W. 1937. Bird Studies at Old Cape May. Vol. I-II. Delaware Valley Ornithol. Club, Philadelphia.

U.S. Dept. of Commerce, Weather Bureau. 1965a. Daily Weather Map (20 August - 30 November). Washington, D.C.

U.S. Dept. of Commerce, Weather Bureau. 1965b. Local Climatological Data, Cape Hatteras, North Carolina (issued monthly). Washington, D.C.

U.S. Fish and Wildlife Service. 1965. Operation Recovery summary, 1965. Mimeo. report. Migratory Non-Game Bird Studies, Patuxent Wildlife Research Center, Laurel, Md.

Sykes, P. W., Jr. 1967. The fall migration of land birds along the Bodie Island-Pea Island region of the Outer Banks of northeastern North Carolina. Unpubl. M.S. thesis, North Carolina State Univ., Raleigh.

Willoughby, J. E. 1951a. Birds at Cape Hatteras. Atlantic Nat. 6:159-161.

Willoughby, J. E. 1951b. Bird notes from Cape Hatteras. Chat 15:74.

APPENDICES

Appendix A
Orders and Families of Birds Recorded in 1965

Order Falconiformes
- Family Cathartidae
- Family Accipitridae
- Family Falconidae

Order Columbiformes
- Family Columbidae

Order Cuculiformes
- Family Cuculidae

Order Strigiformes
- Family Tytonidae
- Family Strigidae

Order Caprimulgiformes
- Family Caprimulgidae

Order Apodiformes
- Family Apodidae
- Family Trochilidae

Order Piciformes
- Family Picidae

Order Passeriformes
- Family Tyrannidae
- Family Hirundinidae
- Family Corvidae
- Family Paridae
- Family Sittidae
- Family Certhiidae
- Family Troglodytidae
- Family Muscicapidae
- Family Mimidae
- Family Motacillidae
- Family Bombycillidae
- Family Laniidae
- Family Sturnidae
- Family Vireonidae
- Family Emberizidae
- Family Fringillidae

Appendix B
List of Birds Recorded in the Study Area in 1965

Species	Method Recorded				
	Censuing	Banding	Collecting	Other[a]	Total Individuals
Turkey Vulture (*Cathartes aura*)	X				1
Osprey (*Pandion haliaetus*)	X				61
Bald Eagle (*Haliaeetus leucocephalus*)	X				1
Northern Harrier (*Circus cyaneus*)	X				244
Sharp-shinned Hawk (*Accipiter striatus*)	X	X			70
Cooper's Hawk (*Accipiter cooperii*)	X				1
Red-shouldered Hawk (*Buteo lineatus*)	X				33
Swainson's Hawk (*Buteo swainsoni*)	X				1
Red-tailed Hawk (*Buteo jamaicensis*)	X				2
American Kestrel (*Falco sparverius*)	X				459
Merlin (*Falco columbarius*)	X				71
Peregrine Falcon (*Falco peregrinus*)	X				24
Mourning Dove (*Zenaida macroura*)	X				1,200
Black-billed Cuckoo (*Coccyzus erythropthalmus*)	X				8
Yellow-billed Cuckoo (*Coccyzus americanus*)	X	X			34
Common Barn-Owl (*Tyto alba*)	X				18
Short-eared Owl (*Asio flammeus*)	X				3
Northern Saw-whet Owl (*Aegolius acadicus*)			X		1
Common Nighthawk (*Chordeiles minor*)	X				4
Chimney Swift (*Chaetura pelagica*)	X				20
Ruby-throated Hummingbird (*Archilochus colubris*)	X				18
Red-headed Woodpecker (*Melanerpes erythrocephalus*)	X				10
Red-bellied Woodpecker (*Melanerpes carolinus*)				X	1
Yellow-bellied Sapsucker (*Sphyrapicus varius*)	X	X			52
Downy Woodpecker (*Picoides pubescens*)	X	X			200
Hairy Woodpecker (*Picoides villosus*)	X				29
Northern Flicker (*Colaptes auratus*)	X	X			1,504
Olive-sided Flycatcher (*Contopus borealis*)	X				3
Eastern Wood-Pewee (*Contopus virens*)	X	X	X		28
Yellow-bellied Flycatcher (*Empidonax flaviventris*)	X		X		6
Acadian Flycatcher (*Empidonax virescens*)	X	X			2
Least Flycatcher (*Empidonax minimus*)			X		1
Empidonax species	X				43
Eastern Phoebe (*Sayornis phoebe*)	X	X			58
Great Crested Flaycatcher (*Myiarchus crinitus*)	X				61
Western Kingbird (*Tyrannus verticalis*)	X				10
Eastern Kingbird (*Tyrannus tyrannus*)	X				599
Purple Martin (*Progne subis*)	X				102
Tree Swallow (*Tachycineta bicolor*)	X				45,467
Northern Rough-winged Swallow (*Stelgidopteryx serripennis*)	X				1

Appendix B. (continued)

Species	Method Recorded: Censuing	Banding	Collecting	Other[a]	Total Individuals
Bank Swallow (*Riparia riparia*)	X				84
Cliff Swallow (*Hirundo pyrrhonota*)	X				3
Barn Swallow (*Hirundo rustica*)	X				1,378
American Crow (*Corvus brachyrhynchos*)	X				54
Fish Crow (*Corvus ossifragus*)	X				1,769
Carolina Chickadee (*Parus carolinensis*)	X	X			98
Red-breasted Nuthatch (*Sitta canadensis*)	X	X			151
White-breasted Nuthatch (*Sitta carolinensis*)	X	X			4
Brown Creeper (*Certhia americana*)	X	X			227
Carolina Wren (*Thryothorus ludovicianus*)	X	X			810
House Wren (*Troglodytes aedon*)	X	X			186
Winter Wren (*Troglodytes troglodytes*)	X	X			7
Sedge Wren (*Cistothorus platensis*)	X				59
Marsh Wren (*Cistothorus palustris*)	X				39
Golden-crowned Kinglet (*Regulus satrapa*)	X	X			197
Ruby-crowned Kinglet (*Regulus calendula*)	X	X			295
Blue-gray Gnatcatcher (*Polioptila caerulea*)	X				5
Veery (*Catharus fuscescens*)	X				1
Gray-cheeked Thrush (*Catharus minimus*)	X	X			14
Swainson's Thrush (*Catharus ustulatus*)	X	X	X		67
Hermit Thrush (*Catharus guttatus*)	X	X			54
Wood Thrush (*Hylocichla mustelina*)	X	X			4
American Robin (*Turdus migratorius*)	X				560
Gray Catbird (*Dumetella carolinensis*)	X	X			2,132
Northern Mockingbird (*Mimus polyglottos*)	X	X			1,324
Sage Thrasher (*Oreoscoptes montanus*)	X				1
Brown Thrasher (*Toxostoma rufum*)	X	X			617
Water Pipit (*Anthus spinoletta*)	X				90
Sprague's Pipit (*Anthus spragueii*)	X				1
Cedar Waxwing (*Bombycilla cedrorum*)	X				196
Loggerhead Shrike (*Lanius ludovicianus*)	X				5
European Starling (*Sturnus vulgaris*)	X				3,994
White-eyed Vireo (*Vireo griseus*)	X	X			68
Solitary Vireo (*Vireo solitarius*)	X				2
Philadelphia Vireo (*Vireo philadelphicus*)	X	X			2
Red-eyed Vireo (*Vireo olivaceus*)	X	X			21
Blue-winged Warbler (*Vermivora pinus*)	X			X	2
Tennessee Warbler (*Vermivora peregrina*)	X		X		10
Orange-crowned Warbler (*Vermivora celata*)	X	X			6
Nashville Warbler (*Vermivora ruficapilla*)	X				7

Appendix B. (continued)

	Method Recorded				
Species	Censuing	Banding	Collecting	Other[a]	Total Individuals
Northern Parula (*Parula americana*)	X	X			19
Yellow Warbler (*Dendroica petechia*)	X				35
Chestnut-sided Warbler (*Dendroica pensylvanica*)	X				2
Magnolia Warbler (*Dendroica magnolia*)	X	X			54
Cape May Warbler (*Dendroica tigrina*)	X	X			136
Black-throated Blue Warbler (*Dendroica caerulescens*)	X	X	X		43
Yellow-rumped Warbler (*Dendroica coronata*)	X	X			12,313
Black-throated Green Warbler (*Dendroica virens*)	X	X			11
Blackburnian Warbler (*Dendroica fusca*)	X	X			4
Yellow-throated Warbler (*Dendroica dominica*)	X				3
Pine Warbler (*Dendroica pinus*)	X	X			8
Prairie Warbler (*Dendroica discolor*)	X	X			148
Palm Warbler (*Dendroica palmarum*)	X	X			738
Bay-breasted Warbler (*Dendroica castanea*)	X				4
Blackpoll Warbler (*Dendroica striata*)	X	X			95
Black-and-white Warbler (*Mniotilta varia*)	X	X			68
American Redstart (*Setophaga ruticilla*)	X	X			559
Prothonotary Warbler (*Protonotaria citrea*)	X				5
Ovenbird (*Seiurus aurocapillus*)	X	X			7
Northern Waterthrush (*Seiurus noveboracensis*)	X	X			39
Connecticut Warbler (*Oporornis agilis*)	X	X			4
Common Yellowthroat (*Geothlypis trichas*)	X	X	X		1,424
Hooded Warbler (*Wilsonia citrina*)	X				1
Wilson's Warbler (*Wilsonia pusilla*)	X	X	X		4
Canada Warbler (*Wilsonia canadensis*)	X				2
Yellow-breasted Chat (*Icteria virens*)	X	X			11
Summer Tanager (*Piranga rubra*)	X				2
Scarlet Tanager (*Piranga olivacea*)	X	X			11
Northern Cardinal (*Cardinalis cardinalis*)	X	X			196
Rose-breasted Grosbeak (*Pheucticus ludovicianus*)	X	X			9
Blue Grosbeak (*Guiraca caerulea*)	X				11
Indigo Bunting (*Passerina cyanea*)	X	X			35
Dickcissel (*Spiza americana*)	X				3
Rufous-sided Towhee (*Pipilo erythrophthalmus*)	X	X			1,341
Bachman's Sparrow (*Aimophila aestivalis*)	X				1
American Tree Sparrow (*Spizella arborea*)	X				6
Chipping Sparrow (*Spizella passerina*)	X	X			100
Clay-colored Sparrow (*Spizella pallida*)	X				9
Field Sparrow (*Spizella pusilla*)	X	X			432
Vesper Sparrow (*Pooecetes gramineus*)	X				37

Appendix B. (continued)

Species	Method Recorded				
	Censusing	Banding	Collecting	Other[a]	Total Individuals
Lark Sparrow (*Chondestes grammacus*)	X				63
Savannah Sparrow (*Passerculus sandwichensis*)	X				1,738
Ipswich race (*P. s. princeps*)	X				1
Grasshopper Sparrow (*Ammodramus savannarum*)	X				14
Sharp-tailed Sparrow (*Ammodramus caudacutus*)	X				12
Seaside Sparrow (*Ammodramus maritimus*)	X				5
Fox Sparrow (*Passerella iliaca*)	X				9
Song Sparrow (*Melospiza melodia*)	X	X			1,128
Lincoln's Sparrow (*Melospiza lincolnii*)	X			X	8
Swamp Sparrow (*Melospiza georgiana*)	X	X			495
White-throated Sparrow (*Zonotrichia albicollis*)	X	X			384
White-crowned Sparrow (*Zonotrichia leucophrys*)	X		X		206
Dark-eyed Junco (*Junco hyemalis*)	X	X			2,232
Snow Bunting (*Plectrophenax nivalis*)	X				25
Bobolink (*Dolichonyx oryzivorus*)	X				192
Red-winged Blackbird (*Agelaius phoeniceus*)	X	X			13,712
Eastern Meadowlark (*Sturnella magna*)	X				1,251
Western Meadowlark (*Sturnella neglecta*)	X				1
Rusty Blackbird (*Euphagus carolinus*)	X				7
Brewer's Blackbird (*Euphagus cyanocephalus*)	X				1
Boat-tailed Grackle (*Quiscalus major*)	X				4,582
Common Grackle (*Quiscalus quiscula*)	X				2
Brown-headed Cowbird (*Molothrus ater*)	X				481
Northern Oriole (*Icterus galbula*)	X	X			604
Purple Finch (*Carpodacus purpureus*)	X				8
Common Redpoll (*Carduelis flammea*)	X				1
Pine Siskin (*Carduelis pinus*)	X				46
American Goldfinch (*Carduelis tristis*)	X				67
Evening Grosbeak (*Coccothraustes vespertinus*)	X				3
House Sparrow (*Passer domesticus*)	X				279
TOTAL					110,482

[a]Includes birds recorded, but not caught, during mist-netting and banding.

Appendix C
Ranking of Species by the Total Number of Individuals Recorded in 1965 by Censuing, Banding, and Collecting

Rank	Species	Number of Individuals
1.	Tree Swallow	45,467
2.	Red-winged Blackbird	13,712
3.	Yellow-rumped Warbler	12,313
4.	Boat-tailed Grackle	4,582
5.	European Starling	3,994
6.	Dark-eyed Junco	2,232
7.	Gray Catbird	2,132
8.	Fish Crow	1,769
9.	Savannah Sparrow	1,738
10.	Northern Flicker	1,504
11.	Common Yellowthroat	1,424
12.	Barn Swallow	1,378
13.	Rufous-sided Towhee	1,341
14.	Northern Mockingbird	1,324
15.	Eastern Meadowlark	1,251
16.	Mourning Dove	1,200
17.	Song Sparrow	1,128
18.	Carolina Wren	810
19.	Palm Warbler	738
20.	Brown Thrasher	617
21.	Northern Oriole	604
22.	Eastern Kingbird	599
23.	American Robin	560
24.	American Redstart	559
25.	Swamp Sparrow	495
26.	Brown-headed Cowbird	481
27.	American Kestrel	459
28.	Field Sparrow	432
29.	White-throated Sparrow	384
30.	Ruby-crowned Kinglet	295
31.	House Sparrow	279
32.	Northern Harrier	244
33.	Brown Creeper	227
34.	White-crowned Sparrow	206
35.	Downy Woodpecker	200
36.	Golden-crowned Kinglet	197
37.	Cedar Waxwing	196
37.	Northern Cardinal	196
38.	Bobolink	192
39.	House Wren	186
40.	Red-breasted Nuthatch	151
41.	Prairie Warbler	148
42.	Cape May Warbler	136
43.	Purple Martin	102
44.	Chipping Sparrow	100
45.	Carolina Chickadee	98
46.	Blackpoll Warbler	95
47.	Water Pipit	90
48.	Bank Swallow	84
49.	Merlin	71
50.	Sharp-shinned Hawk	70
51.	White-eyed Vireo	68
51.	Black-and-white Warbler	68
52.	Swainson's Thrush	67
52.	American Goldfinch	67
53.	Lark Sparrow	63
54.	Osprey	61
54.	Great Crested Flycatcher	61
55.	Sedge Wren	59
56.	Eastern Phoebe	58
57.	American Crow	54
57.	Hermit Thrush	54
57.	Magnolia Warbler	54
58.	Yellow-bellied Sapsucker	52
59.	Pine Siskin	46
60.	Black-thr. Blue Warbler	43
61.	Marsh Wren	39
61.	Northern Waterthrush	39
62.	Vesper Sparrow	37
63.	Yellow Warbler	35
63.	Indigo Bunting	35
64.	Yellow-billed Cuckoo	34
65.	Red-shouldered Hawk	33
66.	Hairy Woodpecker	29
67.	Eastern Wood-Pewee	28
68.	Snow Bunting	25
69.	Peregrine Falcon	24
70.	Red-eyed Vireo	21
71.	Chimney Swift	20
72.	Northern Parula	19
73.	Common Barn-Owl	18
73.	Ruby-thr. Hummingbird	18
74.	Gray-cheeked Thrush	14
74.	Grasshopper Sparrow	14
75.	Sharp-tailed Sparrow	12
76.	Black-thr. Green Warbler	11
76.	Yellow-breasted Chat	11
76.	Blue Grosbeak	11
76.	Scarlet Tanager	11
77.	Red-headed Woodpecker	10
77.	Western Kingbird	10
77.	Tennessee Warbler	10
78.	Rose-breasted Grosbeak	9
78.	Clay-colored Sparrow	9
78.	Fox Sparrow	9
79.	Black-billed Cuckoo	8
79.	Pine Warbler	8
79.	Lincoln's Sparrow	8
79.	Purple Finch	8
80.	Winter Wren	7

Appendix C. (continued)

Rank	Species	Number of Individuals
80.	Nashville Warbler	7
80.	Ovenbird	7
80.	Rusty Blackbird	7
81.	Yellow-bellied Flycatcher	6
81.	Orange-crowned Warbler	6
81.	American Tree Sparrow	6
82.	Blue-gray Gnatcatcher	5
82.	Loggerhead Shrike	5
82.	Prothonotary Warbler	5
82.	Seaside Sparrow	5
83.	Common Nighthawk	4
83.	White-breasted Nuthatch	4
83.	Wood Thrush	4
83.	Blackburnian Warbler	4
83.	Bay-breasted Warbler	4
83.	Connecticut Warbler	4
83.	Wilson's Warbler	4
84.	Short-eared Owl	3
84.	Olive-sided Flycatcher	3
84.	Cliff Swallow	3
84.	Yellow-throated Warbler	3
84.	Dickcissel	3
84.	Evening Grosbeak	3
85.	Red-tailed Hawk	2
85.	Acadian Flycatcher	2
85.	Solitary Vireo	2
85.	Philadelphia Vireo	2
85.	Blue-winged Warbler	2
85.	Chestnut-sided Warbler	2
85.	Canada Warbler	2
85.	Common Grackle	2
85.	Summer Tanager	2
86.	Turkey Vulture	1
86.	Bald Eagle	1
86.	Cooper's Hawk	1
86.	Swainson's Hawk	1
86.	Northern Saw-whet Owl	1
86.	Red-bellied Woodpecker	1
86.	Least Flycatcher	1
86.	N. Rough-winged Swallow	1
86.	Veery	1
86.	Sage Thrasher	1
86.	Sprague's Pipit	1
86.	Hooded Warbler	1
86.	Bachman's Sparrow	1
86.	Savannah (Ipswich) Sparrow	1
86.	Western Meadowlark	1
86.	Brewer's Blackbird	1
86.	Common Redpoll	1

Appendix D
Documentation of Accidental Species Recorded in the Study Area During 1965

SWAINSON'S HAWK

A Swainson's Hawk was found on the north end of Pea Island, and details were published in *The Chat* (Potter and Sykes 1980). The account is as follows:

"Sykes (1967) recorded a Swainson's Hawk in Dare County on 11 October 1965 during a study of landbird migration on the Outer Banks. At 1450 the bird was found perched on the top of a fence post at the intersection of NC 12 and the road to the Oregon Inlet Coast Guard Station on the northern tip of Pea Island. The hawk was studied at 9 m with 7 x 35 binoculars under clear sky conditions. When the bird was at rest, heavy brownish streaks formed an almost solid band across the breast and faded gradually into the belly. The throat and belly were white. The head was streaked with brown, and the back was rather uniform dark brown. The dorsal surface of the tail was brownish-gray with four or five narrow dark bands and a narrow white terminal band. It was not noted whether or not the last tail band was wider than the others. The bird appeared slender for a buteo. The tips of the folded wings extended almost to the tip of the tail. When the hawk flushed, it flew away from the observer a short distance at a gradual climb to gain altitude, at which time a buffy or whitish patch was seen in the rump area. The hawk then soared overhead for a short time before it moved off and disappeared. While the bird was soaring, the tips of the wings appeared distinctly pointed and the broad wings formed a dihedral as in the Marsh Hawk (*Circus cyaneus*). When viewed overhead, the light wing linings and the belly contrasted with the dark primaries and secondaries, and the underside of the tail appeared rather light. This individual was obviously an immature Swainson's Hawk."

SAGE THRASHER

A Sage Thrasher was seen at Whalebone Junction several hundred meters northeast of my Whalebone census area about 1300 hours on 5 October 1965. It was found at an old concrete slab (flooring of a building long gone) that was partly covered with wind-blown sand, scattered grasses and herbaceous plants, and bits of debris. It was observed for about 5 minutes at 6 meters (20 feet) as it hopped about foraging for insects. When the bird flushed, it flew a short distance and alighted on the roof of a low building. Because the bird was on private property, I went to obtain permission to collect it; but when I returned about 15 minutes later, it could not be relocated. The thrasher appeared slightly smaller than the Northern Mockingbird with grayish-brown upperparts and tail, light gray or whitish superciliary line, and buffy auricular region, flanks, and undertail coverts. The underparts were mostly white with distinct dark streaks, which were heaviest on the breast. The undertail coverts were finely streaked. There were two white wing bars, the upper being somewhat indistinct. The tail was short, rounded at the tip, and fanned when the bird flew; the tips of the outer rectrices were white. The bill was short, relatively straight, and dark in color; the legs were also dark. The eyes were light yellow. The plumage of the bird appeared to be in good condition. This bird is believed to have been an adult based on eye color; immatures have dark eyes.

SPRAGUE'S PIPIT

A Sprague's Pipit was flushed from a plowed field on the west side of N.C. 12 just south of North Pond on Pea Island in midafternoon of 30 November 1965. It alighted on the side of a sand dune 2 meters (7 feet) from the east side of the highway and was studied for about a minute. I approached to within 11 m (35 feet) of the bird and had it in clear view. The pipit then flew into the vegetated beach dunes and was not seen again. It did not appear to be associating with any other birds. It was rather light in overall coloration, much lighter than a Water Pipit. The upperparts were light brown with streaks on the back. The underparts were white with a buffy wash on the sides and flanks and fine dark streaks on the chest. The cheeks were a very light buff, the eyes were dark, and the legs a dark flesh color. The outer rectrices were white. I did not see this bird bob its tail, as is the habit of the Water Pipit, or hear it utter a call note. During the brief observation I did not note the bill color.

WESTERN MEADOWLARK

The Western Meadowlark was identified on the basis of its song, which was the typical primary song of the species. Lanyon (1957), in his study of the meadowlarks in Wisconsin, found that occasionally adult males of both species in areas where the breeding ranges meet have been known to sing the song of either species. The call note of the meadowlark on Pea Island was not heard. The call notes of the Eastern Meadowlark (*S. magna*) and Western Meadowlark are quite different. The call notes are innate, whereas the songs are learned (Lanyon 1957). With these reservations in mind, the bird was identified as a Western Meadowlark. Several Eastern Meadowlarks were singing at the time, affording good comparison. The western bird was heard several times, but was never located visually among the numerous eastern birds.

BREWER'S BLACKBIRD

An adult male Brewer's Blackbird, the first for North Carolina's barrier beaches in Dare and Hyde Counties, was seen feeding on the shoulder of the loop road on the west side of the Visitors Center at the Bodie Island Lighthouse in direct sunlight during late morning of 14 November 1965. The bird was about the size of a male Red-winged Blackbird, the plumage was black with a distinct purple head reflection and much iridescence on the body feathers, and the eyes were yellow. The legs were black as was the typical blackbird-shaped bill.

CONTRIBUTIONS OF THE NORTH CAROLINA BIOLOGICAL SURVEY AND THE NORTH CAROLINA STATE MUSEUM—1984

Shelley, Rowland M. Revision of the milliped genus *Dynoria*(Polydesmida: Xystodesmidae). Proceedings of the Biological Society of Washington 97:90-102. 1984-1

Grant, Gilbert S., Charles V. Paganelli, and Herman Rahn. Microclimate of Gull-billed Tern and Black Skimmer nests. Condor 86:337-338. 1984-2

Shelley, Rowland M. Revision of the milleped genus *Cheiropus* (Polydesmida: Xystodesmidae). Proceedings of the Biological Society of Washington 97:263-284. 1984-3

Shelley, Rowland M. A revision of the milliped genus *Dicellarius* with a revalidation of the genus *Thrinaxoria* (Polydesmida: Xystodesmidae). Proceedings of the Biological Society of Washington 97:473-512. 1984-4

Grant, Gilbert S. Studies 19-23, pages 40-41 in Thirty-sixth Winter Bird Population Study, Calvin L. Cink and Roger L. Boyd (editors). American Birds 38:35-63. 1984-5

Lee, David S., and J. Christopher Haney. The genus *Sula* in the Carolinas: an overview of the phenology and distribution of gannets and boobies in the South Atlantic Bight. Chat 28:29-45. 1984-6

Lee, David S., and Eloise F. Potter. Quantifying breeding birds associated with pocosins: an exercise in community definition and myth slaying. (Abstract). 65th Annual Meeting of The Wilson Ornithological Society. 1984-7

Pettit, Ted N., Gilbert S. Grant, and G. Causey Whittow. Nestling metabolism and growth in the Black Noddy and White Tern. Condor 86:83-85. 1984-8

Grant, Gilbert S., and G. Causey Whittow. 1984. Metabolic rate of Laysan Albatross and Bonin Petrel chicks on Midway Atoll. Pacific Science 38:170-176. 1984-9

Lee, David S. Petrels and storm-petrels in North Carolina's offshore waters: including species previously unrecorded for North America. American Birds 38:151-163. 1984-10

Potter, Eloise F. 1984. On capitalization of vernacular names of species. Auk 101:895-896. 1984-11

Grant, G. S. Microclimate of Gull-billed Tern and Black Skimmer nests. (Abstract). 65th Annual Meeting of The Wilson Ornithological Society. 1984-12

Lee, David S. Second supplement to the 1978 Checklist of North Carolina Birds. Chat 48:85-88. 1984-13

Cooper, John E. Vanishing species: the dilemma of resources without price tags. Pages 7-32 in Threatened and Endangered Plants and Animals of Maryland, Arnold W. Norden, Donald C. Forester, and George H. Fenwick (editors). Maryland Natural Heritage Program Special Publication 84-I, Maryland Department of Natural Resources. 475 p. 1984-14

Lee, David S., Steven P. Platania, Arnold W. Norden, Carter R. Gilbert, and Richard Franz. Endangered, threatened, and extirpated freshwater fishes of Maryland. Pages 287-328 in Threatened and Endangered Plants and Animals of Maryland, Arnold W. Norden, Donald C. Forester, and George H. Fenwick (editors). Maryland Natural Heritage Program Special Publication 84-I, Maryland Department of Natural Resources. 475 p. 1984-15

Lee, David S. Maryland's vanished birds and mammals: reflections of ethics past. Pages 454-471 in Threatened and Endangered Plants and Animals of Maryland, Arnold W. Norden, Donald C. Forester, and George H. Fenwick (editors). Maryland Natural Heritage Program Special Publication 84-I, Maryland Department of Natural Resources. 475 p. 1984-16

Braswell, Alvin L., and William M. Palmer. *Cemophora coccinea copei*: Reproduction. Herpetological Review 15(2):49. 1984-17

Braswell, Alvin L. Here come the Frogs! Carolina Notebook 4(4):1-4. 1984-18

SPECIAL PUBLICATIONS
OF THE
NORTH CAROLINA STATE MUSEUM OF NATURAL SCIENCES
AND THE
NORTH CAROLINA BIOLOGICAL SURVEY

Title	Price
Endangered and Threatened Plants and Animals of North Carolina. Cooper, Robinson, and Funderburg, 1977.	Out of print
Atlas of North American Freshwater Fishes. Lee, Gilbert, Hocutt, Jenkins, McAllister, and Stauffer, 1980.	$25, postpaid
Contributions of the North Carolina State Museum of Natural History and the North Carolina Biological Survey, 1884 - 1980.	$ 1, postpaid
A Distributional survey of North Carolina Mammals. Lee, Funderburg, and Clark, 1982.	$ 5, postpaid
Supplement to the Atlas of North American Freshwater Fishes. Lee, Platania, and Burgess, 1983.	$ 5, postpaid
The Seaside Sparrow, Its Biology and Management. Quay, Funderburg, Lee, Potter, and Robbins, 1983.	$15, postpaid
Autumn Land-bird Migration on the Barrier Islands of Northeastern North Carolina. Sykes, 1986.	$ 5, postpaid

Order from: North Carolina State Museum of Natural Sciences
P.O. Box 27647
Raleigh, N.C. 27611
(Make checks payable to NCDA Museum Extension Fund.)

A Division of the North Carolina Department of Agriculture
James A. Graham, Commissioner